# INCOMPRESSIBLE
# FLOW TURBOMACHINES

# INCOMPRESSIBLE FLOW TURBOMACHINES
## Design, Selection, Applications, and Theory

**George F. Round**

*Professor Emeritus*
*McMaster University*
*Hamilton, Ontario*
*Canada*

ELSEVIER
BUTTERWORTH
HEINEMANN

Amsterdam   Boston   Heidelberg   London   New York   Oxford   Paris
San Diego   San Francisco   Singapore   Sydney   Tokyo

Elsevier Butterworth–Heinemann
200 Wheeler Road, Burlington, MA 01803, USA
Linacre House, Jordan Hill, Oxford OX2 8DP, UK

**Library of Congress Cataloging-in-Publication Data**
Application submitted.


**British Library Cataloguing-in-Publication Data**
A catalogue record for this book is available from the British Library.

ISBN: 0-7506-7603-5

For information on all Butterworth–Heinemann publications
visit our Web site at www.bh.com

04   05   06   07   08   09   10      10  9  8  7  6  5  4  3  2  1
Printed in the United States of America

*To Amanda, Christopher, Gillian, and Rachel—who are all very fine*

# CONTENTS

# 4 *Pumps,* 103

## 5  *Some Aspects of Design,* 188

# 9  Cavitation,  269

# 10  Water Hammer,  281

# 11  Corrosion,  299

# Preface

The principal objective of this textbook is to present the basic theory and design of turbomachines, together with applications of such machines. In this context, the term *turbomachine* refers to rotational machines—both turbines and pumps that have blades or impellers. It excludes all reciprocating machines and rotational machines such as screw-type pumps and gear pumps. Nevertheless, the coverage is wide, and it should serve as a useful text for students taking final courses in fluid mechanics and graduate students pursuing research work on turbomachines. It should also be useful as a reference and refresher tool for industry, as well as a reference book for those people involved in day-to-day work in this area. It is not intended to be a design handbook as such, although some design problems are treated in some detail. In this regard, it should be used as a supplement to the excellent handbooks that are available.

The literature in this field is voluminous, so it is with some trepidation that I am adding to it. But most of the literature is scattered, and books in the field are generally concerned with specific segments of turbomachinery, such as pumps, gas turbines and hydraulic turbines, fans, compressors, hydraulic drives, couplings, and the like. In addition, they usually do not present solved problems to illustrate the theory and empirical data. Therefore, it is hoped that in this regard the book will fill a suitable niche in the literature.

Some of the difficulties that have arisen for both designers and users of turbomachines have resulted from inconsistencies in nomenclature and the use of *dimensionless groups* that are not dimensionless. It cannot be overemphasized that care must always be taken to ensure that the nomenclature is consistent when making simultaneous references to texts and to check that any calculations and manipulation of equations are done in a dimensionally consistent way.

The advent of high-speed computing has helped in establishing and confirming designs. Since the earliest days up to the use of high-speed computing, design methods were entirely graphical. These are still very useful and should not be dismissed out of hand. In fact, as the first task the designer should make preliminary designs using computer-aided design tools or computer graphics in order to grasp the problems likely to be encountered. This is particularly true of runners and impellers of double curvature.

Fluid flows through impellers or runners of hydraulic machines are complex, and the theoretical prediction of the performance of a turbine or pump of a given design is a blend of theory and empirical experimental evidence. A large amount of experimental data is available in the literature, and most of it, because of the complexity of the flows, relates to specific machines. This is also true of computer numerical flow software. Ultimately, as is true of all fluid mechanics, the prediction of performance—theoretically, empirically, or numerically for a given set of input variables—must always end with experimental verification.

Acknowledgment of sources of written reference material and diagrams is made throughout the text for individual items. However, I should like to extend my thanks individually to:

VA TECH HYDRO, Zurich, Switzerland, for permission to use various turbine diagrams of projects around the world for which they have been designers.

Sulzer Pumps Ltd., Winterthur, Switzerland for allowing me to use a number of diagrams from their invaluable *Sulzer Centrifugal Pump Handbook* and in particular Fritz Allenbach for his help.

Goulds Pumps of Seneca Falls, New York, in particular Cliff Dodge, for permission to use published catalog material.

John Wiley & Sons, Inc. for granting me permission to reproduce diagrams from the classic book by A.J. Stepanoff *Centrifugal and Axial Flow Pumps*, Copyright © 1957 by John Wiley & Sons. This material is used by permission of John Wiley & Sons, Inc. Acknowledgment is made throughout the text for diagrams.

ASME International for allowing me to reproduce the Moody friction factor diagram.

BHR Solutions for allowing me to reproduce several diagrams from "Internal Flow Systems" by D.S.Miller.

**George F. Round**
McMaster University

# Nomenclature

## English

A = area
b = width
$C_D$ = drag coefficient
$C_L$ = lift coefficient
c = absolute velocity of fluid for pump or turbine
$\mathbf{c}$ = velocity vector, $\mathbf{c} = \mathbf{i}u + \mathbf{j}v + \mathbf{k}w$: $\mathbf{i, j, k}$ = unit vectors
d, D = diameter, drag force
E = modulus of elasticity
f = friction factor
g = acceleration due to gravity
H = net effective head (turbine), head developed by pump
$H_{ATM}$ = atmospheric pressure expressed as a head
$H_G$ = gage pressure expressed as a head
$H_S$ = suction head of turbine or pump
$H_{SV}$ = net positive suction head of turbine or pump
$H_{th}$ = theoretical head developed by a pump with n-blades
$H_{th}(\infty)$ = the head developed by a pump with an infinite number of blades, that is, no circulatory losses
$H_{VAP}$ = vapor pressure expressed as a head
I = second moment of area, mass moment of inertia
L = lift force
M = total mass or moment
N = rotational speed
NPSH = net positive suction head
$N_S$ = specific speed
p = pressure

Q = volumetric flow rate

**R** = universal gas constant

R = outer radius

Re = Reynolds number

r = inner radius

S = suction specific speed

s = pitch

T = temperature

t = time, clearance distance

u = peripheral or tip velocity, x-component of velocity

$V_{av}$ = average fluid velocity

v = y-component of velocity

w = relative velocity, z-component of velocity

z = number of blades or vanes

# Greek

$\alpha$ = angle between c and U

$\beta$ = angle between w and U

$\Gamma$ = circulation

$\gamma$ = specific weight

$\zeta$ = vorticity

$\eta$ = efficiency

$\theta$ = general symbol for angle

$\mu$ = slip factor

$\rho$ = density

$\sigma$ = cavitation parameter

$\tau$ = torque

$\nu$ = kinematic viscosity

$\Phi$ = flow coefficient

$\varphi$ = central angle

$\Psi$ = head coefficient

$\omega$ = angular velocity

# Dimensions of Fluid Mechanics Quantities

| Quantity | Dimensions | Units—SI |
|---|---|---|
| Absolute viscosity | $M\,L^{-1}\,t^{-1}$ | $kg\,m^{-1}\,s^{-1}$ |
| Acceleration | $L\,t^{-2}$ | $m\,s^{-2}$ |
| Area | $L^2$ | $m^2$ |
| Circulation | $L^2\,t^{-1}$ | $m^2\,s^{-1}$ |
| Enthalpy (unit) | $L^2\,t^{-2}$ | $J\,kg^{-1}$ |
| Entropy | $M\,L^2\,t^{-2}\,T^{-1}$ | $J\,K^{-1}$ |
| Force | $M\,L\,t^{-2}$ | N |
| Gas constant | $L^2\,t^{-2}\,T^{-1}$ | $J\,kg^{-1}\,K^{-1}$ |
| Kinematic viscosity | $L^2\,t^{-1}$ | $m^2\,s^{-1}$ |
| Length | L | m |
| Mass | M | kg |
| Mass flow rate | $M\,t^{-1}$ | $kg\,s^{-1}$ |
| Mass moment of inertia | $M\,L^2$ | $kg\,m^2$ |
| Momentum | $M\,L\,t^{-1}$ | $kg\,m\,s^{-1}$ |
| Pressure | $M\,L^{-1}\,t^{-2}$ | $N\,m^{-2}$ |
| Shear stress | $M\,L^{-1}\,t^{-2}$ | $N\,m^{-2}$ |
| Strain rate | $t^{-1}$ | $s^{-1}$ |
| Stream function | $L^2\,t^{-1}$ | $m^2\,s^{-1}$ |
| Temperature | T | K |
| Thermal conductivity | $M\,L\,t^{-3}\,T^{-1}$ | $J\,m^{-1}\,s^{-1}\,K^{-1}$ |
| Torque | $M\,L^2\,t^{-2}$ | N m |
| Velocity | $L\,t^{-1}$ | $m\,s^{-1}$ |
| Velocity potential | $L^2\,t^{-1}$ | $m^2\,s^{-1}$ |
| Volume | $L^3$ | $m^3$ |
| Volumetric flow rate | $L^3\,t^{-1}$ | $m^3\,s^{-1}$ |
| Vorticity | $t^{-1}$ | $rad\,s^{-1}$ |
| Weight | $M\,L\,t^{-2}$ | N |
| Work | $M\,L^2\,t^{-2}$ | J |

Dimensions: mass—M; length—L; time—t; temperature—T

# Units

## SI (Système Internationale) Units

Mass—kilogram (kg)
Length—meter (m)
Time—second (s)
Temperature—degrees Kelvin (K)

## Derived units

Force—Newton (N)
Energy—Joule (J)
Power—Watt (W)
Pressure—Pascal (Pa)

# Fundamental Definitions

*Biot-Savart law*—At any point in a flow field, the presence of a vortex causes an increase in velocity. This is the hydrodynamic analogy of the original Biot-Savart law that specifies the magnetic field strength induced by an electric conductor.

*Circulation*—vorticity per unit area, that is, $d\Gamma = d\zeta /dA$
  In Cartesian coordinates:

$$\Gamma =_S (u\ dx + v\ dy + w\ dz)$$

*Compound vortex*—a combination of a free vortex and a forced vortex.

## Efficiencies—Turbines

### Hydraulic Efficiency

$\eta_H$ = effective head across turbine/actual head across turbine = $(H - H_L)/H$

$H_L$ = total head losses of the turbine

### Mechanical Efficiency

$\eta_M$ = (Power generated)/(Power generated + Losses) = $P/[(\rho g)(Q - Q_L)(H - H_L)]$

### Volumetric Efficiency

$$\eta_V = (Q - Q_L)/Q$$

Usually, volumetric efficiency is 96 to 100%. In many cases it is ignored.

## Overall Efficiency

$$\eta_O = \eta_H H \eta_M H \eta_V = [(H - H_L)/H]HP/[(\rho g)(Q - Q_L)(H - H_L)]H[(Q - Q_L)/Q]$$

$$= P/[(\rho g)(Q)(H)]$$

P = power generated by the turbine;   $[(\rho g)(Q)(H)]$ = input hydraulic energy

Typically, the units are: P = watts (N-m/s); Q = m$^3$/s; H = m; g = m/s$^2$

# Efficiencies—Pumps

## Hydraulic Efficiency

$$\eta_{HYD} = \text{head developed by pump/head delivered to liquid} = H/(u_2 c_{U2} - u_1 c_{U1})/g$$

For pumps, usually $u_1 c_{U1} = 0$, i.e., no pre-whirl

$\eta_{HYD}$ may also be written as: $H/(H + H_L)$

## Mechanical Efficiency

$\eta_M$ = power delivered by pump/power supplied to shaft = $(P - P_L)/P$
Sometimes written as: $\eta_M = (P_{SHAFT} - P_{FRICTION})/P_{SHAFT}$

## Volumetric Efficiency

$$\eta_V = Q/(Q + Q_L)$$

Q = delivered volumetric flow rate

## Overall Efficiency

$$\eta_O = \eta_H H \eta_M H \eta_V = [(\rho g)(Q)(H)]/P$$

$[(\rho g)(Q)(H)]$ = energy possessed by the exit fluid; P = energy input into pump

*Free vortex*—In free vortex flow, the tangential velocity component $c_\theta$ is inversely proportional to the distance from the center of the vortex, r; that is, $c_\theta r$ = constant

*Forced vortex*—In forced vortex flow, the tangential velocity component $c_\theta$ is directly proportional to the distance from the center of the vortex, r, that is, $c_\theta = \omega r$

## Heads

*Actual head across turbine, H*—head difference across the turbine not taking into account losses in the machine

*Theoretical head* for pump with infinite number of blades, $H_{th}$ (4)

*Theoretical head* for pump with a finite number of blades, $H_{th}$ (z)

*Actual pump head, H*—head generated by the pump in height of fluid units, that is, meters or feet of liquid equivalent to the difference in head between inlet (suction) and outlet (discharge)

*Suction head, $H_S$*—head of liquid between the surface (pressured or unpressured) of the liquid being pumped and the centerline of the pump on the suction side

*Discharge head, $H_D$*—head of liquid between the surface (pressured or unpressured) of the liquid being pumped and the centerline of the pump on the discharge side

*Head loss in turbine or pump, $H_L$*—difference in the head delivered by the turbine or pump and the head delivered to the fluid by the impeller or to the runner

*Meridional velocity, $c_M$*—velocity of fluid particles passing through meridional planes passing through the axis of the runner or impeller.

## Power

*Turbine hydraulic power*—the power possessed by the liquid prior to entry to the runner

*Pump hydraulic power*—the power possessed by the liquid at pump exit:

$$P = \gamma H_M Q$$

Q represents the real discharge from the pump outlet

*Rotation*—the angular rotation of the liquid

*Streamline*—the path that a liquid particle takes at steady state through the liquid

*Velocity potential*—a scalar function which, when differentiated with respect to space, gives the velocity. Thus:

$$u = (\delta\Phi/\delta x); \; v = (\delta\Phi/\delta y); \; w = (\delta\Phi/\delta z)$$

*Vorticity*—$\zeta = 2\Omega$, where $\Omega$ = angular velocity of rotation of a fluid at any point
  Note that $\zeta$ and $\Omega$ have three unit vector components.
  Thus:

$$\zeta = i\zeta_x + j\zeta_y + k\zeta_z$$

# CHAPTER 1

# HISTORICAL BACKGROUND AND PRESENT STATE OF DEVELOPMENT

## 1.1 Greek and Roman Machines

Turbomachines notably the Archimedean screw and the Hero "engine" have come down to us in various modified forms from the earliest times to the present day. The invention of the Archimedean screw is usually attributed to Archimedes of Alexandria (287–212 B.C.), and it is thought that the idea came to him on a visit to Egypt. However, the basic principle of the Archimedean screw probably existed in a more primitive form in ancient Egypt. This type of screw, known to the Romans as the Cocleon, was used for raising water from the Nile. Vitruvius has given us the only surviving description of the screw from antiquity (Simms, 1995). He stated that the angle of the screw should be 37° for optimal performance and that the maximum angle should be 45°.

Men walking on cleats fixed around the outer shell of the screw supplied the rotational force needed to operate the first Archimedean screws. Similarly, for the second machine some evidence suggests it is doubtful that the actual invention of the Hero engine or Aeoliphile, the forerunner of the steam turbine, can be directly attributed to Hero of Alexandria (sometime after 150 B.C.). However, Hero was a prolific inventor, and his inventions included a water-powered organ, water-powered automatic door openers, presses for grapes and olive oil, a screw-cutting machine, and a crossbow. All of these inventions are described in his work *Pneumatica*.

Figure 1-1 shows the type and layout of Archimedean screws used in ancient mines. The Romans used such systems for pumping water out of nonferrous mines, that is, lead and copper mines and the like. In the German Museum in Munich is an engraving of a set of Archimedean screws pumping water from a lead mine in A.D. 200 (Klemm, 1959). The water was pumped from a depth of 200 m, with each screw being able to lift water through a height of 1.5 m. (Figure 1-2 shows schematically a form of Cocleon or Egyptian screw, and Figure 1-3 shows a Hero engine.)

The Archimedean or Egyptian screw is a progressive cavity device. Although the main uses were originally for raising water (and remain so today in the Middle East), these devices are also

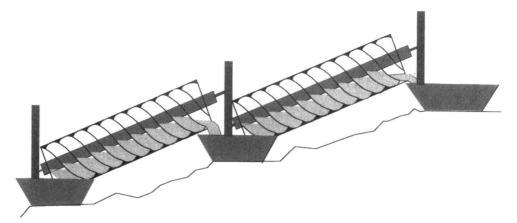

**Figure 1-1** Typical Archimedean screws for raising water from mines by the ancient Romans.

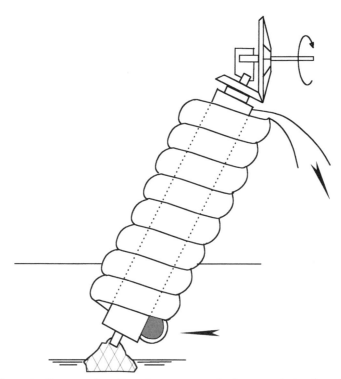

**Figure 1-2** Schematic diagram of an Egyptian screw or Cocleon—a device for raising water for irrigation.

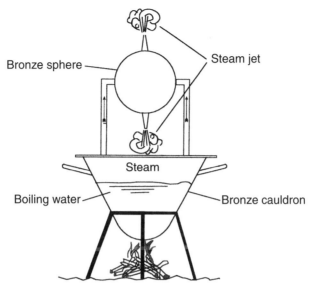

**Figure 1-3** Hero engine or Aeoliphile.

used to move solid powdery or granular material, such as grains and fibers. The material flows or is loaded by hopper into the rotating entrance and progresses sequentially by a series of steps. The steps are, in effect, the moving chambers. In Egyptian and Roman times, the prime movers originally were humans or animals.

The Hero engine is a steam turbine. Although by modern standards, it is an impractical, highly inefficient device, it does demonstrate the main principles of a modern steam turbine. The spherical drum shown in Figure 1-3 contains water and is heated by a wood fire underneath. The water boils, turning into steam which issues forth as jets located symmetrically on each side of the drum. The ensuing rate of change of momentum causes a force to act in the opposite direction on each jet. The resultant torque causes the drum to rotate.

The Romans' principal concern with regard to hydraulic machines was with moving water from one location to another—water supply systems. This was done either vertically using different forms of Archimedean or Egyptian screws or horizontally using aqueducts or pipes. Stopcocks and faucets were not used, and water flowed continuously. Reservoirs or stilling basins were used if the water came from several sources. The two most famous engineers of the Roman era were Marcus Vitruvius Pollio, who was in charge of the water supply during the rebuilding of Rome by Octavian, and Sextus Julius Frontinius who was commisioner of water for Rome in A.D. 97. Frontinius's treatise, *De Aquis Urbis Romae*, gives a complete description of water distribution methods in Rome. Some of the surviving aqueducts today are directly attributable to Vitruvius and Frontinius.

Vitruvius's description in his *De Architectura Libri Decem* shows that the undershot waterwheel was in use as early 25 B.C. The overshot waterwheel was known in late Roman times.

By the early fourth century A.D., the Romans had built a mill for flour production near Arles, fed by an aqueduct (Williams, 1987). Two sets of eight overshot waterwheels on each side of the mill provided the mill with about 22 kW of power.

## 1.2 The Middle Ages

The fall of the western part of the Roman Empire in about 400 A.D. was followed by a transition period of about one thousand years—a period frequently called the Dark Ages. Advances in hydraulics that had occurred under such notables as Archimedes effectively ended. Europe became fragmented into small states, and no significant advance in hydraulics and turbomachines occurred. In the Arab world, however, knowledge of waterwheels, Archimedean screws, and water supply systems using combinations of waterwheel and Archimedean screw flourished. Waterwheels originally introduced into Europe by the Romans were improved in Syria during this period, and generally the Middle East developed many ingenious combinations of screws and undershot and overshot waterwheels (Sarton, 1931). In contrast, in the Western world the structures that had been built by the Romans were allowed to deteriorate.

At the beginning of the fourteenth century Europe had furnaces with large air-bellows driven by waterwheels. Figure 1-4 shows an overshot waterwheel for raising water from mines (Agricola, 1556). However, progress in hydraulics and hydraulic machines was still painfully slow because of the fallacious beliefs and arguments that originated from the reigning philosophy of science.

## 1.3 The Renaissance

The second half of the fifteenth century was marked by a number of events that created the atmosphere for rethinking philosophical science and the birth of the experimental method. The resistance to ecclesiastical control of learning was given impetus by the fall of Constantinople in 1453, which gave rise to an influx of scholars from the East to the West, the rediscovery of America, and perhaps most importantly, the development of printing. All of these occurrences permitted more rapid dissemination of knowledge.

One individual was preeminent in the Renaissance years—Leonardo da Vinci (1452–1519). Leonardo excelled at everything he touched; it is probably fair to say that he was the greatest genius the world has ever seen. His talents ranged from painting, music, natural philosophy, anatomy, and botany to mechanics, engineering, and architecture. From 1502 to 1503, he served as chief engineer to Cesare Borgia, designing military weapons and equipment and supervising the design of canals and harbors. The principle of continuity might justifiably be credited to Leonardo, together with the idea of using centrifugal force for lifting liquids. His treatise on hydraulics (Carusi and Favaro, 1924) was in itself a monumental work. Leonardo had also thought about hydraulic motors and deduced correctly that overshot waterwheels were intrinsically more efficient than undershot wheels.

**Figure 1-4** Overshot water wheel for raising water from mines—from a woodcut in *De re metallica* by Agricola (1556).

# 1.4 The Post Renaissance

The first centrifugal pump was built more than 15 centuries ago; it was made of wood, and the impeller had double curvature. Although Leonardo first suggested using centrifugal force for lifting water, the invention of the predecessor of the modern centrifugal pump has been attributed to Denis Papin (1647–1714). The rotor of this pump had two flat radial blades rotating in a closed cylindrical casing.

Early in the 1700s minor improvements were made on pump design, but one obstacle to the development of centrifugal pumps was the popularity of piston-type pumps. This was because centrifugal pumps require fairly high-speed drives, and certainly at the beginning of centrifugal pump development these drives were not available. However, major improvements were made on waterwheel design both for vertical undershot and overshot wheels and for horizontal wheels. John Smeaton made a series of experiments on model waterwheels in 1752 and communicated his results to the Royal Society in 1759. Among his results, he noted that overshot wheels were twice as efficient as undershot wheels, and the impact of streams of water on a flat plate showed a marked loss of energy in the form of spray and turbulence. Jean-Charles made the first analysis and design of horizontal waterwheels in 1767. Figure 1-5 shows a horizontal waterwheel designed by Borda, another pioneer of this period.

Jean Victor Poncelet designed and built another very successful waterwheel in the early nineteenth century (Smith, 1977). An adjustable sluice controlled the flow through this undershot

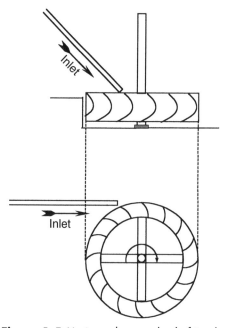

**Figure 1-5** Horizontal waterwheel of Borda.

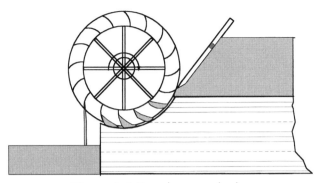

**Figure 1-6** Poncelet waterwheel.

wheel. Figure 1-6 is a representation of a Poncelet wheel. Although these wheels were primitive, pressureless turbines—that is, the available pressure energy was converted to kinetic energy before entering the blades—these wheels set the scene and were the forerunners of the work that later enabled reaction tubines to be designed and successfully operated.

## 1.5 The Nineteenth Century to the Present

The first hydraulic machine, having recognizable elements in modern-day turbines, was devised by Claude Burdin (1790–1873). The machine that Burdin designed was an efflux type and was entered in a national competition in France. Although Burdin did not win the competition, he was awarded a consolation prize and encouraged to continue his research. Unfortunately, Burdin never built a prototype or even a working model. The design also lacked a casing, the clearances were not sufficiently small, and the runner blades were not well shaped in the aerodynamic or hydrodynamic sense. An illustration of the Burdin design is shown in Figure 1-7.

The problems of the Burdin design were eliminated in the turbine design of Benoit Fourneyron (1802–1867). Fourneyron was a student of Burdin at L' Ecole des Mines in Saint-Etienne. After leaving the Ecole, between 1823 and 1827, he developed and tested his first experimental water turbine, a free-efflux type, at Pont-sur-l'Ognon.

Fourneyron's results also involved the first successful application of the Prony brake as a dynamometer. By bringing together all the previous ideas on waterwheel and turbine technology, a successful machine could now be built. In 1833, Fourneyron entered the same competition as Burdin had entered, and he won it. After that, Fourneyron successfully designed and built over 100 turbines for installations in Europe and around the world. Industrial development in the latter half of the eighteenth century was beginning to get underway, and efficiency considerations now became of paramount importance.

An excellent review of waterwheel development in the eighteenth and nineteenth centuries leading to and containing elements of the modern hydraulic turbine has been made by Smith (1977).

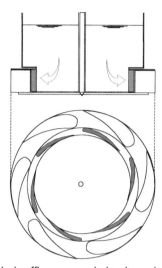

**Figure 1-7** Six-blade efflux-type radial turbine designed by Burdin.

A short time after the success of the Fourneyron machine, in 1844, a company in the United States acquired the rights to a turbine designed by Uriah Boyden. The chief engineer, James Francis (1815–1892), was given the task of improving its performance. Present-day Francis turbines of the inward radial-flow type are due to his pioneering work. During the nineteenth century, a number of impulse turbines were also invented. Only one has survived to the present day—the Pelton wheel—named in honor of Lester A. Pelton (1829–1908). In 1880, Pelton was granted a U.S. patent on a vertical waterwheel around the periphery of which were flat plates on which a jet of water from a nozzle impinged. Tests at the University of California at that time showed that the efficiency of the machine was about 40%. In 1889, Pelton patented an improved bucket construction, together with an improved means of bucket adjustment. Figure 1-8 is taken from Pelton's 1889 patent showing the improved bucket construction and means of adjustment. It may be seen that the jet is split with the bowls inclined. The modern form of bucket is ellipsoidal in shape.

The nineteenth century is also noteworthy because of a dramatic increase in experimental work in other areas of hydraulics such as open channel flows (St. Venant, Dupuit) and groundwater flows (Darcy), measuring devices (Pitot), and towing tanks (Froude). In addition, the first systematic investigation of rotodynamic pumps was made in the 1890s at the Sultzer Brothers factory in Switzerland, and after this, development was rapid.

The nineteenth and twentieth centuries were marked by great strides toward effecting a better fundamental understanding of the flow of fluids—hydrodynamics. Although the basic equations of fluid mechanics have never been solved in their general form, a number of specific solutions have been made. High-speed computers have also enabled specific numerical solutions for problems that hitherto have remained intractable; such advances will continue at a rapid rate.

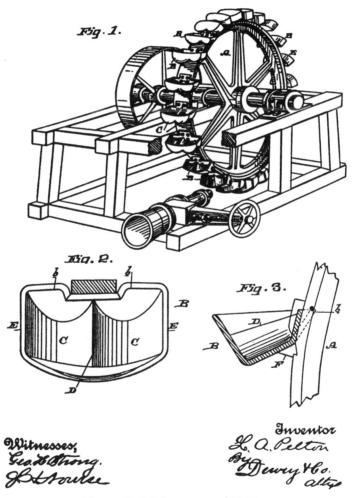

**Figure 1-8** Pelton patent of 1889.

# 1.6 General Classification of Rotodynamic Turbines and Pumps

Incompressible fluid turbines extract energy from a flowing fluid, almost invariably water, in a manner that is both efficient and environmentally benign. The means by which this is done influences the path of the fluid through the machine, and the path in turn is influenced by the design of the machine. The type of machine design that is most suitable is determined by the conditions governing the fluid flow, such as head or pressure across the machine and the volumetric flow rate of fluid through it. Thus, the spectrum of flow conditions covers flows that vary from high heads and low-flow rates to low heads and high-flow rates. The flows vary from high-speed jets at relatively low-flow rates to medium to high-flow rates. Turbines are classified generally in this way. An important and useful parameter for the classification is the specific speed, $N_S$. Specific speed will be discussed in detail in Chapter 3.

Incompressible fluid pumps increase the pressure energy of the fluid in an efficient manner. Again, as with turbines, the means by which this is done influences the path of the fluid through the machine, and the path in turn influences the design of the machine. Pumps considered here are rotary dynamic and displacement dynamic, with the rotary dynamic pump being in greater preponderance. Rotary dynamic pumps are classified according to impeller shape. The pumps considered here have rotating shafts about which variously configured blades or impellers are attached. These cause a flow through the pump that varies from purely radial flow to purely axial flow. Displacement dynamic pumps move a fluid from one location, the inlet, to another location, the outlet, in discrete volumes. At the same time, compression increases the pressure in the fluid.

# 1.7 Theoretical Limitations

There are two flow criteria for the efficiency of hydraulic turbines and centrifugal pumps and, for that matter, any turbomachine:

1. The amount of fluid energy transformed by the machine to rotational mechanical energy for turbines and, in the case of pumps, the energy supplied to the shaft compared to the total amount of energy transformed by the pump.
2. The energy distribution across the flow leaving passages in the machine. Lack of uniformity in energy distribution is harmful in the sense that energy is wasted in redistribution or equalization, so that streams of fluid entering and leaving become uniformly energy-distributed between reservoirs.

Modern hydraulic turbines are remarkably efficient, but with large-scale turbines generating thousands of kilowatts, even an efficiency improvement of 1or 2% is worthwhile.

The theory of turbomachinery is based on steady flows. In this way, theoretical solutions may be made as simple as possible. This is not likely to change in the near future, and with high-speed, large-memory computers it probably will not be necessary to delineate the complete flow field in unsteady flow. It should be emphasized that we are concerned not with turbulent fluctuations but rather with the time-dependent variations in relative flow between moving and stationary surfaces.

At present, there is no complete theoretical model of three-dimensional flows. The assumption of frictionless fluids to make the problems tractable means that it is doubtful whether the flow field can represent the real phenomena taking place. In the forefront of present-day problems are the design of turbine micro- and mini-systems, boiler feed pumps for operation at supercritical pressures, multistage reversible pumps, pumps for use with molten metals, heavy water, radioactive liquids, very low temperature liquids, and so on.

# 1.8 References

Carusi, E., and Favaro, A. (Eds.). *Leonardo da Vinci, Del Moto e Misura dell'Acqua.* Bologna (1924).

Herschel, C., *The Two Books on the Water Supply of Rome of Sextus Julius Frontinius.* New York (1913).

Klemm, F., *A History of Western Technology.* Allen and Unwin Ltd., London (1959).

Sarton, G., *Introduction to the History of Science.* Vol. II, Part II, Carnegie Institution of Washington (1931).

Simms, D.L., *Archimedes the Engineer, History of Technology.* Vol. 17, Mansell Information Publishing Ltd., London (1995).

Smith, N.A.F., *The Water Turbine and Its Name. History of Technology.* Vol. 2, Mansell Information Publishing Ltd., London (1977).

Williams, T.I., *The History of Invention.* Facts on File Publications, New York (1987).

# THEORY OF TURBOMACHINES

## 2.1 Equations Governing the Behavior of Turbomachines

All the equations describing the fluid dynamic behavior of turbomachines are conservation equations:

1. Conservation of mass flow rate—continuity equation
2. Conservation of linear momentum—integral momentum equation
3. Conservation of angular momentum—angular momentum equation
4. The Euler turbine equation
5. Conservation of energy—the Bernoulli equation
6. General steady-flow energy equation

## 2.2 Continuity Equation

The continuity equation as a mass conservation equation implies that for a given control volume, a balance exists between the masses of fluid entering and the masses leaving per unit time and the change in density. The equation may be written in differential form as:

$$(\partial\rho/\partial t) + \nabla \cdot (\rho c) = (D/Dt) + \rho\nabla \cdot c = 0 \tag{2.1}$$

The velocity vector is:

$$c = \mathbf{i}u + \mathbf{j}v + \mathbf{k}w \tag{2.2}$$

For steady-flow incompressible fluid (density = constant), Equation (2.1) reduces to: $\nabla \cdot \mathbf{c} = 0$

  (D/Dt) is the substantive derivative, composed of a local contribution due to unsteady flow and a convective contribution due to translation.

Equation (2.1) in indicial notation may be written:

$$(\partial\rho/\partial t) + \partial(\rho c_i)/\partial x_i = (D\rho/Dt) + \rho(\partial c_i/\partial x_i) = 0 \tag{2.3}$$

## 2.3 Linear Momentum Theorem

One of the most important aspects of calculating the change of any property of a large fluid body is the identification of the system in space and time, so that physical laws may be applied to its behavior. This is a relatively simple thing to do with a solid body, but fluid systems are not so easily identifiable because they tend to deform to occupy space available and continuously deform when in motion. The easiest way to do this with fluids is to observe the behavior of fluid elements as they pass through an identifiable volume.

The basic relationship that enables us to do this is the Reynolds Transport Theorem:

$$DN/Dt = \int_{CS} n(\rho\mathbf{c} \cdot dA) + (\partial/\partial t) \int_{CV} n\rho\, dV \tag{2.4}$$

N is an extensive property, and n is the corresponding intensive property in Equation (2.4).

When linear momentum is taken as the extensive property, $n = c$, and Equation (2.4) becomes:

$$F = \int_{CS} \mathbf{c}(\rho\mathbf{c} \cdot dA) + (\partial/\partial t) \int_{CV} \mathbf{c}\rho\, dV \tag{2.5}$$

F is the resultant force on the system due to all the surface and body forces, $F_S$ and $F_B$. Hence:

$$F_S + F_B = \int_{CS} \mathbf{c}(\rho\mathbf{c} \cdot dA) + (\partial/\partial t) \int_{CV} \mathbf{c}\rho\, dV \tag{2.6}$$

As an example of the use of Equation (2.6), we may use it to evaluate the forces acting on a reducing elbow by a fluid entering with velocity $c_1$ and exiting with velocity $c_2$. Figure 2-1 illustrates the problem.

Referring to Figure 2-1, we may break down the components of Equation (2.6) into:

$$(F_S + F_B)_X = p_1A_1 - p_2A_2\cos\beta + F_X \tag{2.7}$$

$$(F_S + F_B)_Y = -p_2A_2\sin\beta - mg + F_Y \tag{2.8}$$

If the flow is steady, the second term on the right-hand side of Equation (2.4) is zero and the first term becomes:

$$\int_{CS} \mathbf{c}(\rho\mathbf{c} \cdot dA) = (dm/dt)(c_2 - c_1) = (dm/dt)[(c_2\cos\beta - c_1) + c_2\sin\beta] \tag{2.9}$$

where $(dm/dt) = \rho_1A_1c_1 = \rho_2A_2c_2 =$ the mass flow rate.

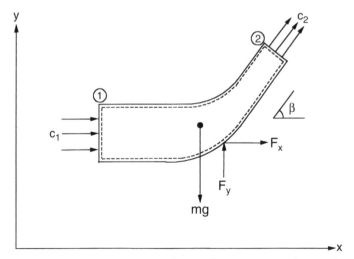

**Figure 2-1** Forces acting on a reducing elbow containing flowing fluid.

Substituting the result into the x- and y-components of $(F_S + F_B)$:

$$F_X = (dm/dt)(c_2\cos\beta - c_1) - p_1A_1 + p_2A_2\cos\beta \qquad (2.10)$$

$$F_Y = (dm/dt)(c_2\sin\beta) + mg + p_2A_2\sin\beta \qquad (2.11)$$

where $p_1$ and $p_2$ are the pressure normal to the faces at inlet and outlet. The inlet pressure is known, and the outlet pressure may be determined from the Bernoulli equation.

## 2.4 Angular Momentum Equation

A fundamental equation for rotating fluid bodies is the angular momentum equation. Consider the two-dimensional fluid element as shown in Figure 2-2. The differential force $dF_t$ may be written:

$$dF_t = \partial(c_t\,dm)/\partial t + (c_t\rho c_n\,dA) \qquad (2.12)$$

where $c_t$ is the tangential component of velocity at radius r.

The moment exerted of the differential element $dT_S$ is given by:

$$dT_S = rd\,dF_t \qquad (2.13)$$

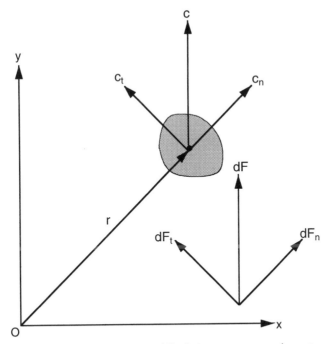

**Figure 2-2** Two-dimensional fluid element rotating about O.

Combining Equations (2.12) and (2.13) gives the equation for the differential torque:

$$dT_S = r\partial(c_t \, dm)/\partial t + (c_t \rho c_n \, dA) \tag{2.14}$$

Integration of Equation (2.14) yields:

$$T_S = \frac{\partial}{\partial t} \iiint rc_t \, dm + \iint rc_t(\rho c_n \, dA) \tag{2.15}$$

Equation (2.15) can be generalized into a vector form applicable to a system in three dimensions.

$$T_S = \frac{\partial}{\partial t} \iiint (r \times c) \, dm + \iint (r \times c)(\rho c \cdot dA) \tag{2.16}$$

## 2.5 Euler Turbine Equation

In 1784, Leonhard Euler formulated an angular momentum balance equation for a flowing liquid; this is directly applicable to a fluid passing through a turbine. If a runner rotates at constant angular velocity due to a liquid stream passing through it, then the torque produced on the runner is equal to the rate of change of angular momentum.

$$M = (\rho Q)(r_1 c_1 \cos \alpha_1 - r_2 c_2 \cos \alpha_2) \tag{2.17}$$

Equation (2.17) applies equally well to pumps, with the modification that the flow is reversed so that the notation of Equation (2.17) is also reversed, that is,

$$M = (\rho Q)(r_2 c_2 \cos \alpha_2 - r_1 c_1 \cos \alpha_1) \tag{2.18}$$

For turbines, the power transmitted to the shaft is:

$$P = M = (\rho Q)(r_1 c_{u1} - r_2 c_{u2})\omega \tag{2.19}$$

For pumps, the power transmitted to the liquid is:

$$P = M = (\rho Q)(r_2 c_{u2} - r_1 c_{u1})\omega \tag{2.20}$$

For Equations (2.19) and (2.20), $\omega$ = angular rate of rotation. In each case, the subscript 1 denotes fluid entering the machine, and the subscript 2 denotes fluid leaving the machine. Substituting Equation (2.17) into Equation (2.19) and Equation (2.18) into Equation (2.20), we obtain for turbines:

$$P = (\gamma/g)Q(c_1 u_1 \cos \alpha_1 - c_2 u_2 \cos \alpha_2) \tag{2.21}$$

For pumps:

$$P = (\gamma/g)Q(c_2 u_2 \cos \alpha_2 - c_1 u_1 \cos \alpha_1) \tag{2.22}$$

It should be noted that $\alpha$ is denoted differently by different authors. It is also denoted as $(1 - \alpha)$. What should be kept in mind is that we are always talking about the tangential component of the absolute velocity. If another common notation is used, then it will be specified in an accompanying vector diagram.

From the trigonometry of the velocity triangles for turbines and pumps, Equations (2.21) and (2.22) may be written:

$$P = (\gamma Q)[\Delta(c^2)/2g + \Delta(u^2)/2g + \Delta(w^2)/2g] \tag{2.23}$$

where:

$\Delta(c^2)/2g$—represents the change in kinetic energy of the liquid
$\Delta(u^2)/2g$—represents the energy expended in causing circumferential flow
$\Delta(w^2)/2g$—represents the change in relative energy from inlet to outlet

# 2.6 Bernoulli Equation

The Bernoulli equation for the steady flow of an ideal fluid expresses an energy balance between two sections of a fluid connected by the same streamline. It may be written in the form:

$$p_1/\gamma + c_1^2/2g + z_1 = p_2/\gamma + c_2^2/2g + z_2 \tag{2.24}$$

The first term represents pressure energy, the second, kinetic energy, and the third, potential energy. The units are length (e.g., meters). For a real fluid Equation (2.24) must be modified to allow for friction and other losses. It may be written:

$$p_1/\gamma + c_1^2/2g + z_1 = p_2/\gamma + c_2^2/2g + z_2 + h_f + h_{other} \tag{2.25}$$

$h_f$ represents frictional losses, and $h_{other}$ represents other nonfrictional losses.

The unsteady form of the Bernoulli equation for an ideal fluid integrated along a streamline is:

$$p_1/\gamma + c_1^2/2g + z_1 = p_2/\gamma + c_2^2/2g + z_2 + \int (\partial c/\partial t)\, ds \tag{2.26}$$

## 2.6.1 Example: Use of Bernoulli Equation for Radial Flow

As a first step in modeling the complex flow in a centrifugal impeller or runner, we consider the purely radial, steady-state flow of a fluid through an annular gap between two disks as shown in

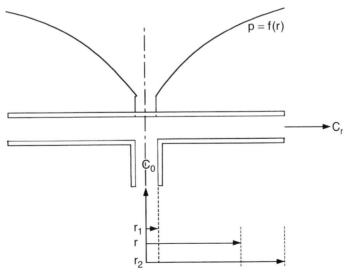

**Figure 2-3** Radial flow through the annular space between two disks.

Figure 2-3. In this example, the application would be for a centrifugal pump. If the volumetric flow rate into the center is Q, then the velocity at any point at radius r from the center is:

$$c = Q/(2\pi rb) \tag{2.27}$$

where:

  $b$ = distance separating the disks

The velocities at positions $r_1$ and $r_2$ are therefore:

$$c_1 = Q/(2\pi r_1 b) \tag{2.28}$$

and

$$c_2 = Q/(2\pi r_2 b) \tag{2.29}$$

Applying the Bernoulli equation between points 1 and 2:

$$p_1/\gamma + c_1^2/2g = p_2/\gamma + c_2^2/2g \tag{2.30}$$

In this instance, $z_1 = z_2$ and viscous losses are neglected (inviscid fluid). Substituting Equations (2.28) and (2.29) in Equation (2.30):

$$p_1/\gamma - p_2/\gamma = c_1^2/2g - c_2^2/2g = Q^2/(8\pi^2 b^2 g)[1/r_1^2 - 1/r_2^2] \tag{2.31}$$

The curve corresponding to this equation is shown in Figure 2-3 and is known as Barlow's curve.

## 2.7 The Energy Equation

The energy equation is a mathematical statement of the First Law of Thermodynamics, and as such it is also a conservation equation. The total energy per unit mass of a system of particles, whether they are solid or fluid, is: $E/M = e$.

  e is composed of several parts:

$$e = u + \frac{1}{2}c^2 + gz + e_{add} \tag{2.32}$$

where:

  $u$ = internal energy
  $\frac{1}{2}c^2$ = kinetic energy
  $gz$ = potential energy

Energy conservation states that for a system consisting of a mass of particles:

$$de = \delta q - dw \qquad (2.33)$$

where q represents differential heat exchange, $\delta q$ is used of course, because q is a point function, and dw is differential work, a path function. In Equation (2.32) $e_{add}$ includes such effects as electromagnetic fields, chemical reactions, lattice energy, and nuclear reactions. In the study of turbomachines, $e_{add}$ is irrelevant. For a system of particles, Equation (2.33) may be written:

$$E = Q - W \qquad (2.34)$$

Heat Q and work W are not properties of the system but represent energy movement across boundaries in various forms. The term W may consist of shaft work, viscous shear work, and flow work. Thus, a general conservation equation may be written:

$$[d(Q - W)/dt]_{CS} = [(E/t)]_{CV} + \iint (u + c^2 + gz)\rho c_n \, dA \qquad (2.35)$$

With enthalpy defined as:

$$h = u + p/\rho \qquad (2.36)$$

Equation (2.35) becomes:

$$[d(Q - W)/dt]_{CS} = [(E/t)]_{CV} + \iint (u + p/\rho + c^2 + gz)\,\rho c_n \, dA \qquad (2.37)$$

In steady-state flow with an inlet (1) into the system and an outlet (2) from the system, Equation (2.37) becomes:

$$[d(Q - W)/dt]_{CS} = \{[(h + c^2/2 + gz)]_{OUT} - [(h + c^2/2 + gz)]_{IN}\}A \qquad (2.38)$$

## 2.8 Similarity

In order to predict the behavior of a prototype machine from a scaled model and vice versa, two conditions of similarity must be met:

1. They must be geometrically similar.
2. They must be dynamically similar.

Condition 1 is met if the ratio of all homologous, that is, mutually corresponding, sections are the same: ratios of characteristic lengths and angles contained within the sections must be the same. Thus, impeller diameter ratio, shape disposition, and number of impeller blades and guide vanes must be the same. The relative roughness of the solid boundaries in contact with fluid must also be the same. Condition 2 is met if at corresponding points in the machines there are similar velocity vector triangles. Generally speaking, it is impossible to satisfy conditions 1 and 2 together with the requirement that viscous effects must be similar. This means that the machines usually have unequal Reynolds numbers. Experiments have shown that the Reynolds number effect is small and may be ignored most of the time.

## 2.9 Dimensional Analysis

Application of dimensional analysis to geometrically similar turbomachines yields four dimensionless groups:

1. $(gH)/(ND)^2$
2. $P/(N^3D^5)$
3. $(Q/ND^3)$
4. Reynolds no. $Re = (VD/\mu)$

1. $(gH)/(ND)^2$ is referred to as the head coefficient and is given the symbol, $\Psi$.
2. $P/(N^3D^5)$ is called the power coefficient and is given the symbol P.
3. $(Q/ND^3)$ is called the flow coefficient and is given the symbol, $\Phi$.

1 and 2 are both functions of 3; thus:

$$(gH)/(ND)^2 = f[(Q/ND^3)] \tag{2.39}$$

$$P/(N^3D^5) = f[(Q/ND^3)] \tag{2.40}$$

In addition, $\Psi$ and P are both weak functions of Re, and again this group is usually ignored. Equation (2.39) applies to centrifugal pumps, and Equation (2.40) applies to turbines. Arising out of Equations (2.39) and (2.40), we obtain a series of affinity relations:

$$Q_1/Q_2 = (N_1D_1^3)/(N_2D_2^3) \tag{2.41}$$

$$H_1/H_2 = (N_1^2 D_1^2)/(N_2^2 D_2^2) \tag{2.42}$$

$$P_1/P_2 = (N_1^3 D_1^5)/(N_2^3 D_2^5) \tag{2.43}$$

In addition, Equations (2.42) and (2.43) enable the common definition of specific speed $N_S$, for turbines to be obtained:

$$N_S = NP^{0.5}/\rho^{1/2}(gH)^{5/4} \tag{2.44}$$

Similarly, from Equations (2.41) and (2.42) for pumps:

$$N_S = NQ^{0.5}/(gH)^{3/4} \tag{2.45}$$

Consistent units in Equations (2.44) and (2.45) make them dimensionless. It is common practice to omit g and ρ from the Equation (2.44), and it is usually written:

$$N_S = Np^{0.5}/(H)^{5/4} \qquad (2.46)$$

Similarly, with Equation (2.45) it is usual to omit g; thus for pumps:

$$N_S = NQ^{0.5}/(H)^{3/4} \qquad (2.47)$$

Because of the omission of g from Equations (2.46) and (2.47), and ρ from Equation (2.46) they are not dimensionless. Another specific speed in use for pumps is the so-called suction specific speed, defined as:

$$S = NQ^{0.5}/(NPSH)^{5/4} \qquad (2.48)$$

Units in common use for Equations (2.47) and (2.48) are:

| (A) | (B) | (C) | (D) |
|---|---|---|---|
| (rpm, US gpm, ft., bhp) | (rpm, ft³/s, ft., bhp) | (rpm, liters/s, m, kW) | (rpm, m³/s, m, kW) |

## 2.10 Restrictions on Similarity Applications

Any characteristic curve or set of characteristic curves for any turbomachine cannot be determined theoretically. Each curve must be determined by experiment. The use of Equations (2.41) to (2.43) enables the relationship between homologous machines to be determined if the performance of one of the machines is known. There is no theoretical relation between two points on the same curve. Thus, if we have determined experimentally the P-H relation for a model turbine and the prediction of the theoretical behavior of a prototype or another set of operating conditions is needed, then Equation (2.46) is needed. For example, the relation between P and H for a turbine operating at a different N will be given by:

$$P = KH^{5/2} \qquad (2.49)$$

In Equation (2.49), $K = (N_S/N)^2$ and $N_S$ has the same maximum value at each condition.

If we consider the H-Q curve for a pump at a particular set of operating conditions and we wish to determine the H-Q curve for the same N, then this must be done by the use of Equation (2.47). No single-relation theoretical equation will do this. Thus:

$$H = KQ^{2/3} \qquad (2.50)$$

In Equation (2.50), $K = (N/N_S)^{4/3}$ and $N_S$ has the same maximum value at each condition.

The similarity laws may only be applied to homologous machines for which there are experimental data for one of the machines.

## 2.11 Dimensionless Groups and Specific Speed

Theoretically, Equations (2.41) to (2.43) enable the behavior of homologous machines to be predicted and are thus fundamental parameters in the study of all turbomachines. However, the predictions of behavior of prototype turbines and pumps using these groups sometimes show significant departures from experimental measurements. There are different losses for turbines and pumps. Generally speaking, departures from the scaling predictions from the groups represented by Equations (2.41) to (2.43) are due to:

1. Reynolds number effect
2. Manufacturing tolerance changes
3. Clearance differences
4. Surface finish differences
5. Flow effects not accounted for within the groups
6. Measurement errors and installation misalignments

## 2.12 Scaling Discrepancies

A large number of prototype/model equations can be used for scaling both for turbines and pumps to account for the differences between performance predicted by Equations (2.41) to (2.43). A few in common use are as follows (the author of the equation is shown as well):

**Moody:**

$$(1 - \eta)/(1 - \eta_{MODEL}) = (D_{MODEL}/D)^n \tag{2.51}$$

$$(1 - \eta)/(1 - \eta_{MODEL}) = (H_{MODEL}/H)^{0.01}(D_{MODEL}/D)^{0.25} \tag{2.52}$$

Moody originally developed Equation (2.51) for turbines. Field tests showed that the exponent n should be approximately 0.20. Furthermore, Equation (2.51) should only be applied to reaction turbines. The same remark applies to Equation (2.52), which incorporates a head term as well. In some cases with impulse turbines, prototype efficiencies have been found to be lower than model efficiencies. So Equations (2.51) and (2.52) should be restricted to reaction turbines (i.e., Francis type).

The application of Equations (2.51) and (2.52) to pumps is even more problematic, since the relationships between model and prototype turbines are not the same as the relationships between model and prototype pumps.

**Pfleiderer:**

$$(1 - \eta)/(1 - \eta_{MODEL}) = (Re_{MODEL}/Re)^{0.01}(D_{MODEL}/D)^{0.25} \tag{2.53}$$

## Hutton:

$$(1 - \eta)/(1 - \eta_{\text{MODEL}}) = 0.3 + 0.7(\text{Re}_{\text{MODEL}}/\text{Re})^{0.2} \qquad (2.54)$$

Equation (2.54) is recommended for use with Kaplan and propeller-type turbines rather than Equations (2.51) and (2.52).

## Ackeret:

$$(1 - \eta)/(1 - \eta_{\text{MODEL}}) = 0.5[(\text{Re}_{\text{MODEL}}/\text{Re})^{0.2}] \qquad (2.55)$$

Although modeling and scaling of hydraulic turbines are well established and certainly acceptable within engineering requirements, especially if the manufacturer recommends a particular scaling equation for a specific turbine, the same is not true of pumps. When similar scaling equations are applied to pumps, the equations have one thing in common: lack of agreement. They are all empirical equations derived from experimental data. Nixon (1965) presented and reviewed a number of scaling equations and examined their deficiencies, demonstrating that there is little correlation. Superficially, it would seem reasonable to suppose that if the scaling factors are not too large, that is, fairly close to one, then all of the correlations will predict the scaled behavior closely. However, data presented by Nixon and Cairney (1972) do not show this. Figure 2-4 shows

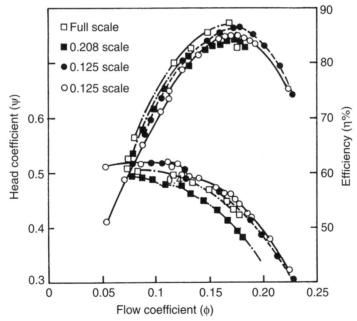

**Figure 2-4** Nondimensional parameters for a specific pump (Eggborough pump) modeled at different scales, together with full scale (adapted from Nixon and Cairney, 1972).

data for several pumps of different geometric scale. The scatter of the data is within $\pm 10\%$, but no direction to the scaling is discernible. Anderson (1977) has suggested that a database from many turbines and pumps be used to estimate probable efficiencies.

## 2.13 Graphical Correlations for Specific Speed

The literature presents a number of generalized graphical correlations for both turbines and pumps. All of these correlations should serve as a guide for selecting a particular machine for a specific function and nothing more. For turbines, parameters such as overall efficiency, head across the machine, and cavitation have been correlated with specific speed. More detailed correlations exist for pumps.

Figure 2-5 shows a generalized correlation for overall efficiency as a function of specific speed, $N_S$ for impulse turbines (Pelton wheels), Francis turbines, and Kaplan-type turbines. It should be noted here that the units of $N_S$ are rpm, kW, and m. These curves indicate the selection that should be made for a particular set of operating conditions. For example, for high heads and low volumetric flow rates, Q, a Pelton wheel is usually the best choice; at the other end of the scale, for low heads and much higher flows, a Kaplan turbine is usually the best choice. In-between Francis-type turbines have a wide range of applicability.

For pumps, a general relationship of head coefficient and of specific speed is shown in Figure 2-6 for a range of different pumps. The shape of a pump impeller has a marked effect on $N_S$ and therefore on performance. Figures 2-7 and 2-8 relate to pumps. Figure 2-7 shows

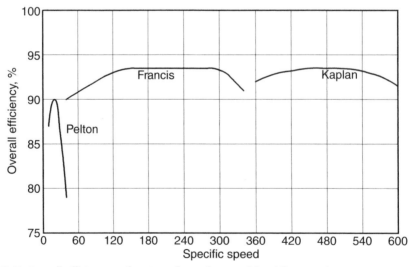

**Figure 2-5** Overall efficiency as function of specific speed for different turbines. (Units of $N_S$-rpm, kW, and m)

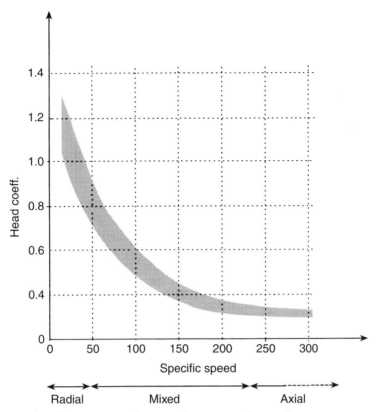

**Figure 2-6** Head coefficient, $\psi$, as a function of pump specific speed, $N_S$. (Units of $N_S$-rpm, $m^3/s$, m.)

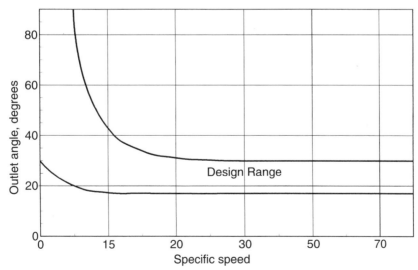

**Figure 2-7** General correlation of $\beta_2$ with $N_S$. (Units of $N_S$-rpm, $m^3/s$, m)

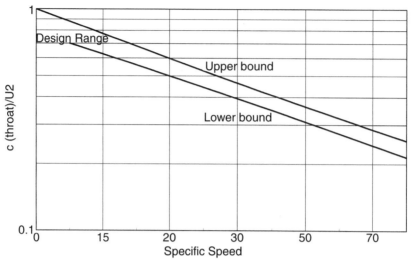

**Figure 2-8** General correlation of $c_{throat}/u_2$ with $N_S$.
(Units of $N_S$-rpm, m³/s, m)

a generalized correlation, useful for design purposes, of the variation of outlet impeller blade angle $\beta_2$ with $N_S$. Figure 2-8 is another generalized correlation showing the variation of $c_{throat}/u_2$ with $N_S$.

## 2.14 General Geometry of Rotational, Radial, and Axial Flows

Outward radial flow is a characteristic of centrifugal pumps or compressors, whereas inward radial flow is a characteristic of Francis-type turbines. The rotational component of the fluid motion may be assumed to follow the equation for constant angular momentum:

$$cr = \text{constant} \tag{2.56}$$

The radial component of the fluid motion, because of continuity, is given by:

$$c_r r = \text{constant} \tag{2.57}$$

The streamline spirals generated by Equations (2.56) and (2.57) are slightly more complicated than logarithmic or Archimedean spirals and generally are not used for impeller vanes in pumps

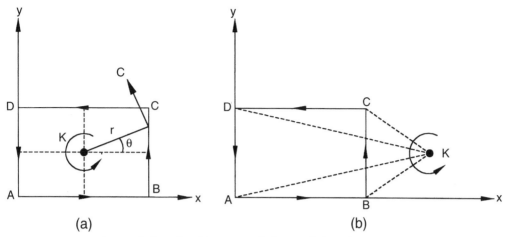

**Figure 2-9** Paths of line integrals for circulation about a vortex.

and runners in turbines. A commonly used form for radial-flow pump impellers is a logarithmic spiral (described in Chapter 4).

## 2.15 Circulation, Free Vortex Flow, and the Kutta-Joukowski Theorem

Circulation is defined as the line integral of the tangential velocity component around a closed contour in a velocity field. Mathematically, circulation is defined as follows:

$$\Gamma = \oint \mathbf{c} \cdot \mathbf{ds}$$

$$\mathbf{c} = \text{velocity vector} \tag{2.58}$$

$$\mathbf{ds} = \text{element of length on the contour}$$

The value of $\Gamma$ depends on the path. Consider a free vortex given by the equation $c_\theta\, r = a$ constant, say K. We consider two paths—one enclosing the center of the vortex and the other not including the vortex (see Figure 2-9). Path (a) encloses the center. The circulation from Equation (2.43) is:

$$\Gamma = \int_A^B c_\theta r\, d\theta + \int_B^C c_\theta r\, d\theta + \int_C^D c_\theta r\, d\theta + \int_D^A c_\theta r\, d\theta \tag{2.59}$$

For a free vortex, $c_\theta r = K$:

$$\Gamma = K \int_A^B d\theta + K \int_B^C d\theta + K \int_C^D d\theta + K \int_D^A d\theta \tag{2.60}$$

Referring to Figure 2-9 (b), clearly the integrals of Equation (2.59) encompass $2\pi$

$$\Gamma = 2\pi K \tag{2.61}$$

That is, for path (a) the circulation is constant.

Similarly, for path (b):

$$\Gamma = \int_A^B c_\theta r \, d\theta + \int_B^C c_\theta r \, d\theta + \int_C^D c_\theta r \, d\theta + \int_D^A c_\theta r \, d\theta \tag{2.62}$$

The path is contained within the angle CKB, but

$$\text{angle CKB} = \text{angle}(AKB + CKD + DK) = K(0) = 0 \tag{2.63}$$

This leads to the important theorem:

"For a flow of constant energy, the circulation around any closed contour not enclosing any force-transmitting body must be zero." This law can be used to show that the circulation around a deflecting body, for given flow conditions, is independent of the size or shape of the contour along which $\Gamma$ is measured. This is illustrated in Figure 2-10. The circulation along the outer contour $C_1$ is:

$$\Gamma_1 = \int_{C_1} c_S = ds \tag{2.64}$$

and along the inner contour $\Gamma_2$. The counterclockwise direction is called positive, and the clockwise direction, negative.

Both $\Gamma_1$ and $\Gamma_2$ are positive. Now consider the circulation around the shaded area between the contours $C_1$ and $C_2$; that is,

$$\Gamma_{1-2} = \Gamma_1 + \oint_{A-B} c_s \cong ds - \Gamma_2 + \oint_{C-D} c_s \cong ds \tag{2.65}$$

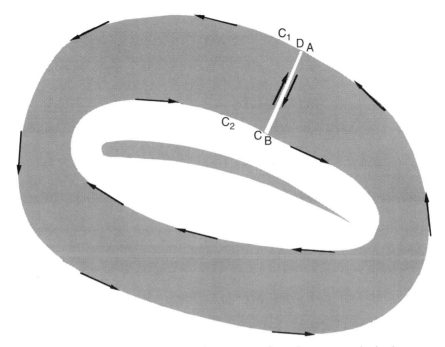

**Figure 2-10** Demonstration of constancy of circulation around a body.

Since A and B coincide and are in opposite directions, it follows that:

$$\oint_{A-B} c_s \cdot ds = \oint_{C-D} c_s \cdot ds \qquad (2.66)$$

that is,

$$\Gamma_{1-2} = \Gamma_1 - \Gamma_2 \qquad (2.67)$$

$\Gamma_{1-2}$ does not enclose any force-transmitting body; therefore, $\Gamma_{1-2} = 0$

$$\therefore \quad \Gamma_1 = \Gamma_2 \qquad (2.68)$$

This means that the integrating contour may be chosen very close to the body. Similarly, it can be shown that the circulation about multiple bodies as in Figure 2-10 is equal to the algebraic sum of the circulations around all parts of the region inside the contour. Referring to Figure 2-11, the paths E-A and D-B have equal magnitude and opposite direction; therefore, they cancel each other. Thus, for the paths indicated:

$$3\oint (EFAE + EABDE + DBGD) = \oint (EFABGDE) = \Gamma_1 + \Gamma_2 + \Gamma_3 \qquad (2.69)$$

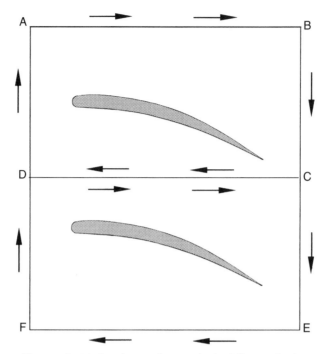

**Figure 2-11** Circulation about multiple deflecting bodies.

It may be seen that an immediate application lies in the field of turbomachinery.

An important theorem relating the lift force experienced by an airfoil or impeller vane to circulation around the airfoil is the Kutta-Joukowski theorem. This theorem was originally derived for circulatory flow around an infinite, circular cylinder and then extended to a cylinder of arbitrary cross section. It may be shown that the lift force L is related to the circulation by means of the equation:

$$L = \rho U_0 \Gamma \tag{2.70}$$

where:

  $\rho$  = fluid density
  $U_0$ = fluid velocity of the free stream approaching the airfoil or impeller vane
  $\Gamma$  = circulation bound to the airfoil

The circulation may or may not satisfy the Kutta condition of smooth flow off the trailing edge of the airfoil. If this condition is entirely met, then this is equivalent to neglecting boundary layer displacement thickness and wake thickness of the trailing edge. The trailing edge condition is illustrated in Figure 2-12.

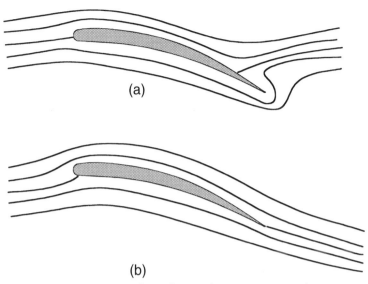

**Figure 2-12** Illustration of the Kutta trailing edge condition: (a) zero circulation, stagnation point on the upper surface; (b) clockwise circulation with a value of ensuring that the stagnation point coincides with the trailing edge.

## 2.16 Forces Acting on an Axial-flow Turbine and Axial-flow Pump Blade

The analysis of flow past an axial blade, whether it is a turbine or pump, is aided by the use of airfoil theory. Initially, we consider a cross section of an isolated airfoil as in Figure 2-13. In this case, the blade is moving fluid with velocity U driven by the fluid; this represents a turbine blade. The velocity triangles at the leading and trailing edges are shown.

A single-velocity vector **w** is chosen to represent the flow past the blade. Its value is given by:

$$\mathbf{w} = (\mathbf{w}_1 + \mathbf{w}_2)/2 \tag{2.71}$$

This parallels the flow along the line making an angle $\theta$ to the chord line, the angle of attack. The direction of $c_a$ is taken to be positive, so that the axial force is:

$$F_A = (L \sin \beta + D \cos \beta) \tag{2.72}$$

The tangential force is:

$$F_T = (L \cos \beta + D \sin \beta) \tag{2.73}$$

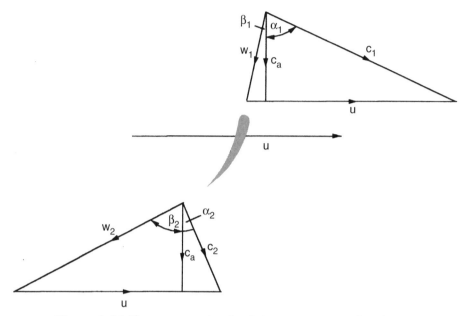

**Figure 2-13** Flow past an isolated airfoil, representing a turbine blade.

The lift force and the drag force are defined in terms of **w** as:

$$L = C_L \, \rho w^2 A_p / 2 \tag{2.74}$$

and

$$D = C_D \, \rho w^2 A_p / 2 \tag{2.75}$$

These quantities are illustrated in Figures 2-14 and 2-15.

$A_p$ is the planform area of the blade and is represented by the product of the blade chord, c, and an incremental length along the blade $\Delta z$. Combining Equations (2.72) and (2.73) with Equations (2.74) and (2.75) yields:

$$F_A = (C_L \, \sin \beta + C_D \, \cos \beta) \, \rho w^2 c \, \Delta z / 2 \tag{2.76}$$

$$F_T = (C_L \, \cos \beta - C_D \, \sin \beta) \, \rho w^2 c \, \Delta z / 2 \tag{2.77}$$

The momentum equation in the tangential direction is written as:

$$F_T = \iint c_t \rho c_n \, dA = \rho c_a (c_{U3} - c_{U2}) 2\pi r \Delta z \tag{2.78}$$

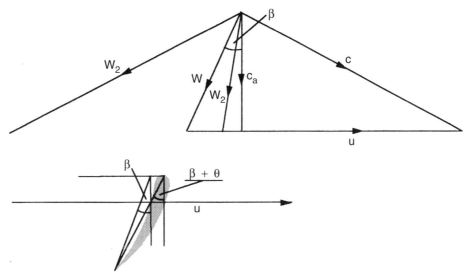

**Figure 2-14** Mean values of relative approach velocity, w, and mean angle, β.

Also:

$$\Sigma F_T = N F_T = N(C_L \cos \beta - C_D \sin \beta) \rho w^2 c \Delta z/2 = \rho c_a(c_{U3} - c_{U2})2\pi r \Delta z \qquad (2.79)$$

Substituting $c_a = w \cos \beta$ yields:

$$(c_{U3} - c_{U2}) = (C_L - C_D \tan \beta)wcN/4\pi r \qquad (2.80)$$

The momentum equation in the axial direction is written as:

$$\Sigma F_A = \iint c_n \rho \, c_n \, dA = \rho c_a(c_{a3} - c_{a2})2\pi r \Delta z \qquad (2.81)$$

$$NF_A = (p_2 - p_3)2\pi r \Delta z$$

Because $c_a = c_{a3} = c_{a2}$, in a similar fashion to Equation (2.73) we may write:

$$N(C_L \sin \beta + C_D \cos \beta) \rho w^2 c \, \Delta z/2 = (p_2 - p_3)2\pi r \Delta z$$

Rearranging:

$$(p_3 - p_2) = -(C_L \sin \beta + C_D \cos \beta) \rho w^2 c \, N/4\pi r \qquad (2.82)$$

In a similar way to the development of the axial-flow turbine equations, there is a corresponding development for equations for an axial-flow pump. Again, a single-velocity vector **w** is chosen to

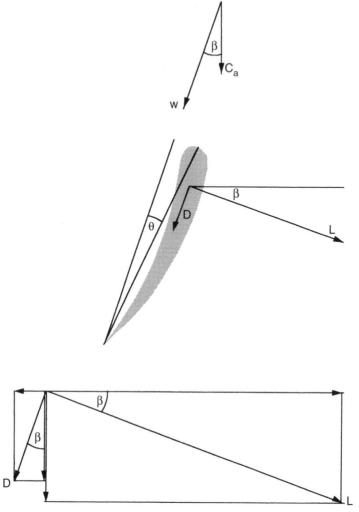

**Figure 2-15** Lift and drag forces on a turbine blade. The components of lift, L, and drag, D, are also shown.

represent the flow past the blade, and similarly to the turbine its value is given by:

$$\mathbf{w} = (\mathbf{w}_1 + \mathbf{w}_2)/2 \tag{2.83}$$

The blade motion and flow are shown in Figure 2-16 together with the velocity triangles. Notice that because this is a pump blade it is inverted. The direction of blade travel remains the same, from left to right.

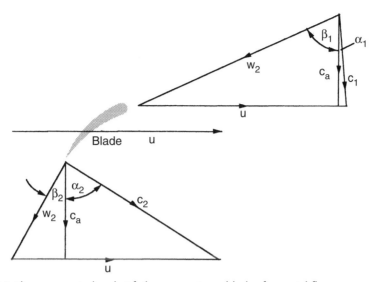

**Figure 2-16** Flow past an isolated airfoil, representing a blade of an axial-flow pump or compressor. The associated velocity triangles at inlet and outlet are shown.

The angle of incidence or attack is designated by $\theta$. The mean relative velocity is given by:

$$\mathbf{w} = c_a/\cos \beta \tag{2.84}$$

The mean flow angle $\beta$ is given by:

$$\tan \beta = (\tan \beta_1 + \tan \beta_2)/2 \tag{2.85}$$

The mean values $\mathbf{w}$ and $\beta$ are shown in Figure 2-17.

The lift, L and drag, D, forces are, respectively, normal to and parallel to the chord line. Again, the chord line is a straight line joining the leading to the trailing edge of the airfoil, as shown in Figure 2-18.

Considering Figure 2-18, the tangential force opposing the direction of motion of the blade is given by:

$$F_T = -(L \cos \beta + D \sin \beta) \tag{2.86}$$

The axial force is given by:

$$F_A = -(L \sin \beta - D \cos \beta) \tag{2.87}$$

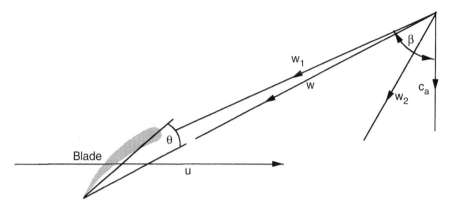

**Figure 2-17** Mean values of relative approach velocity, w, and mean angle, β.

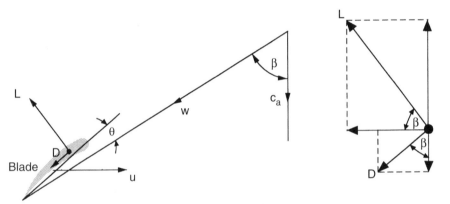

**Figure 2-18** Lift and drag forces on the pump blade. The components of lift, L, and drag, D, are also shown.

Lift and drag forces are usually expressed in terms of lift and drag coefficients given by:

$$L = C_L \rho w^2 A_p / 2 \qquad (2.88)$$

$$D = C_D \rho w^2 A_p / 2 \qquad (2.89)$$

where:

$A_p$ = projected area of the blade normal to the flow = $c\Delta z$

Combining Equations (2.86) and (2.87) with Equations (2.88) and (2.89):

$$F_T = -(C_L \cos \beta + C_D \sin \beta)\rho w^2 c \, \Delta z / 2 \qquad (2.90)$$

$$F_A = -(C_L \sin \beta - C_D \cos \beta)\rho w^2 c \, \Delta z / 2 \qquad (2.91)$$

The momentum equation in the tangential direction is written as:

$$\Sigma F_T = \iint c_t \rho c_n \, dA = \rho c_a (c_{U2} - c_{U1}) 2\pi r \Delta z \qquad (2.92)$$

Setting Equation (2.71) with N blades equal to Equation (2.73), we obtain:

$$N(C_L \cos \beta + C_D \sin \beta)\rho w^2 c \, \Delta z / 2 = \rho c_a (c_{U2} - c_{U1}) 2\pi r \Delta z \qquad (2.93)$$

Substituting $c_a = w \cos \beta$ yields:

$$(c_{U3} - c_{U2}) = (C_L - C_D \tan \beta) w c N / 4\pi r \qquad (2.94)$$

The momentum equation in the axial direction is written as:

$$\Sigma F_A = \iint c_n \rho c_n \, dA = \rho c_a (c_{a2} - c_{a1}) 2\pi r \Delta z \qquad (2.95)$$

$$N F_A = (p_1 - p_2) 2\pi r \Delta z \qquad (2.96)$$

In a similar fashion to Equation (2.74):

$$-N(C_L \sin \beta - C_D \cos \beta)\rho w^2 c \Delta z / 2 = (p_1 - p_2) \, 2\pi r \Delta z \qquad (2.97)$$

Rearranging:

$$(p_1 - p_2) = -(C_L \sin \beta - C_D \cos \beta) 2 \rho w^2 c N / 4\pi r \qquad (2.98)$$

The foregoing equations relate changes in aerodynamic properties across the blades.

## 2.17 Stream Function and Streamlines

Streamlines show the paths taken by fluid elements moving through a fluid in steady flow. The path is everywhere parallel to the local flow; that is, the local velocity vector at any point is tangential to the streamline. The usual symbol for a streamline is $\Psi$. In the special cases of two-dimensional or axisymmetric flows, streamlines can be related to the continuity equation for an incompressible flow:

$$(u/x) + (v/y) = 0 \qquad (2.99)$$

u = velocity in the x-direction
v = velocity in the y-direction

Equation (2.99) is satisfied automatically by a function $\Psi$ known as a stream function:

$$u = (\partial\Psi/\partial y) \qquad\qquad (2.100)$$

$$v = -(\partial\Psi/\partial x) \qquad\qquad (2.101)$$

$\Psi$ is a function of x and y and possibly t.
Thus,

$$d\Psi = -v\,dx + u\,dy \qquad\qquad (2.102)$$

## 2.18 Velocity Potential

Another function $\Phi$—a scalar function, called the velocity potential, which is also a function of x, y, and t—may be introduced. It is defined as:

$$c = \nabla\Phi \qquad\qquad (2.103)$$

$$u = (\Phi/x) \qquad\qquad (2.104)$$

$$v = (\Phi/y) \qquad\qquad (2.105)$$

Thus,

$$(\partial\Phi/\partial x) = (\partial\Psi/\partial y) \qquad\qquad (2.106)$$

$$(\partial\Phi/\partial y) = -(\partial\Psi/\partial x) \qquad\qquad (2.107)$$

Equations (2.96) and (2.97) are known as the Cauchy-Riemann equations. The $\Psi$-lines are streamlines, and the $\Phi$-lines are velocity potential lines in a two-dimensional flow plane. It may easily be shown that:

$$(dy/dx)|_{\Phi=\text{const.}} = -(u/v) \qquad\qquad (2.108)$$

and

$$(dy/dx)|_{\Psi=\text{const.}} = +(v/u) \qquad\qquad (2.109)$$

$$\therefore \quad [(dy/dx)|_{\Phi=\text{const.}}][(dy/dx)|_{\Psi=\text{const.}}] = -1 \qquad\qquad (2.110)$$

Lines of constant $\Psi$ and $\Phi$ are orthogonal; such lines form a flow net. The nodes of such a net have their $\Psi$-lines and $\Phi$-lines intersecting at right angles. Figure 2-19 shows such a flow in a reducing elbow. When the velocity varies, the squares formed by the intersections are curvilinear. This method of graphically describing a flow has direct applicability to the flow in an impeller cavity and the volute of a centrifugal pump.

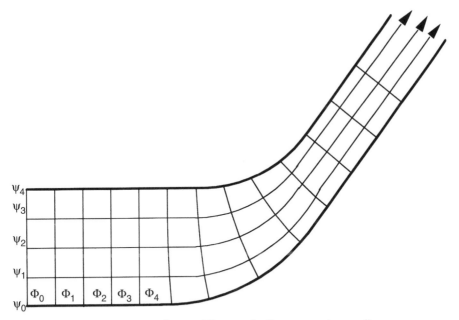

**Figure 2-19** Orthogonal flow net for flow in a reducing elbow.

## 2.19 Superposition of Streamlines

The equations of motion of a steady, two-dimensional, nonviscous flows are:

$$u(\partial u/\partial x) + v(\partial u/\partial y) = (-1/\rho)(\partial p/\partial x) \tag{2.111}$$

$$u(\partial v/\partial x) + v(\partial v/\partial y) = (-1/\rho)(\partial p/\partial y) \tag{2.112}$$

Differentiating Equation (2.111) with respect to y and Equation (2.112) with respect to x, we obtain:

$$(\partial u/\partial y)(\partial u/\partial x) + u(\partial^2 u/\partial y\partial x) + (\partial v/\partial y)(\partial u/\partial y) + v(\partial^2 u/\partial y^2) = (-1/\rho)(\partial^2 p/\partial y\partial x) \tag{2.113}$$

and

$$(\partial u/\partial x)(\partial v/\partial x) + u(\partial^2 v/\partial x^2) + (\partial v/\partial x)(\partial v/\partial y) + v(\partial^2 v/\partial y^2) = (-1/\rho)(\partial^2 p/\partial x\partial y) \tag{2.114}$$

The direction of differentiation is immaterial because the functions are continuous. Therefore, subtracting Equation (2.113) from Equation (2.114) and using the continuity equation, we obtain:

$$u(\partial/\partial x)(\partial v/\partial x - \partial u/\partial y) - v(\partial/\partial y)(\partial v/\partial x - \partial u/\partial y) = 0 \tag{2.115}$$

Equation (2.115) is satisfied if:

$$(\partial v/\partial x - \partial u/\partial y) = 0 \tag{2.116}$$

If we now consider two flows with x- and y-components $u_1$ and $v_1$ and $u_2$ and $v_2$, then both flows satisfy:

$$(\partial u_1/\partial x + \partial v_1/\partial y) = 0 \tag{2.117}$$

$$(\partial u_2/\partial x + \partial v_2/\partial y) = 0 \tag{2.118}$$

$$(\partial v_1/\partial x - \partial u_1/\partial y) = 0 \tag{2.119}$$

$$(\partial v_2/\partial x - \partial u_2/\partial y) = 0 \tag{2.120}$$

These equations are linear so that when they are added algebraically the result is still linear.

$$\therefore \qquad \partial(u_1 + u_2)/\partial x + \partial(v_1 + v_2)/\partial y = 0 \tag{2.121}$$

and

$$\therefore \qquad \partial(v_1 + v_2)/\partial x + \partial(u_1 + u_2)/\partial y = 0 \tag{2.122}$$

The new flow satisfies the same differential equations as the original fluid motions. Figure 2-20 shows graphically the resultant of two such arbitrary flows. Mathematically, the method may be illustrated by the superposition of a radial source flow and a vortex. Thus, the equations for the streamlines and velocity potential lines of a radial source flow are:

$$\Psi = (q/2\pi)\theta \quad \text{and} \quad \Phi = -(q/2\pi)\ln r \tag{2.123}$$

$$q = \text{volumetric flow rate per unit depth}$$

For a vortex:

$$\Psi = (-K/2\pi)\ln r \quad \text{and} \quad \Phi = (-K/2\pi)\ln r \tag{2.124}$$

Adding the $\Psi$ and $\Phi$ functions together algebraically gives the resultant flow field:

$$\Psi = (q/2\pi)\theta - (K/2)\ln r \tag{2.125}$$

$$\Phi = -(q/2\pi)\ln r - (K/2\pi)\ln r \tag{2.126}$$

The result is a spiral approximating the shape of streamlines in the impeller of a centrifugal pump. Figure 2-21 shows the flow pattern. It is immediately apparent that the center of the

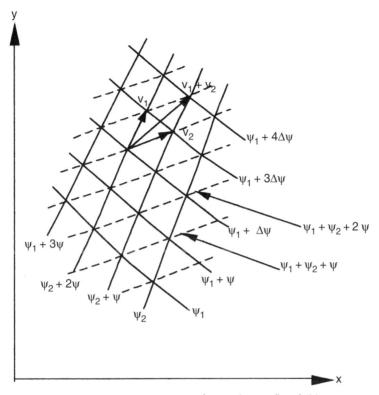

**Figure 2-20** Superposition of two arbitrary flow fields.

spiral is still a singularity because of the line source. However, a better approximation to the impeller pattern is to use the so-called Rankine combined vortex. This vortex is a combination of a core of rotational flow—a forced vortex and an irrotational vortex. A free vortex is illustrated in Figure 2-22. The Rankine combined vortex is shown in Figure 2-23.

## 2.20 Axisymmetric Flows and Stokes's Stream Function

Axially symmetric, three-dimensional flows may be analyzed in a similar way to two-dimensional flows. The axis of symmetry chosen will be the z-axis. On all planes normal to this, the flow characteristics will be a function of r and time, t, and will be independent of the angle, $\theta$. Potential flow applications for turbomachines will be, for example, entrance flows into hubs of turbines and pumps or any other part of a machine that is axially symmetric.

A stream function for this flow is the Stokes's stream function, $\psi$. Consider a point B in the coordinate system shown in Figure 2-24. The surface generated by rotating the point B about

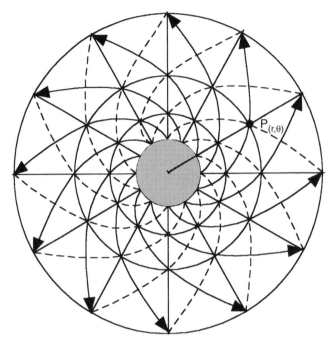

**Figure 2-21** Graphical superposition of a radial source flow and a vortex; the result is a spiral.

the axis of symmetry, z, gives a surface of revolution, the flow through which is independent of the generating curve.

A stream function $\psi$ for the flow in terms of z, R, and time t is:

$$q = 2\pi\psi(z, R, t) \tag{2.127}$$

Note that the same symbol $\psi$ is used for both two- and three-dimensional flows. The volumetric flow through any two points may be expressed as:

$$q_{1-2} = 2\pi(\psi_1 - \psi_2) \tag{2.128}$$

The velocities in the R- and $\beta$-directions are:

$$V_R = [1/(R^2 \sin\beta)](\partial\psi/\partial\beta) \tag{2.129}$$

$$V_\beta = [-1/(R \sin\beta)](\partial\psi/\partial R) \tag{2.130}$$

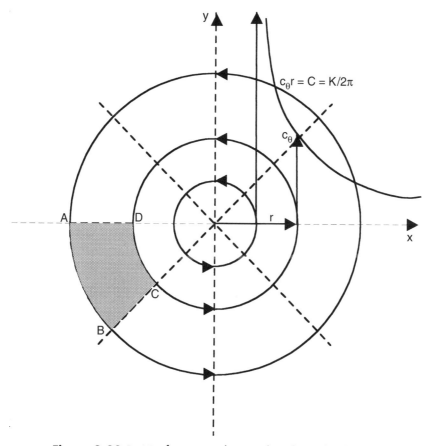

**Figure 2-22** Positive free vortex showing the velocity distribution.

For irrotational flows, the spherical forms of the Cauchy-Riemann equations become:

$$[1/(\sin\beta)](\partial\psi/\partial R) = -(\partial\Phi/\partial\beta) \tag{2.131}$$

$$[1/(R^2\sin\beta)](\partial\psi/\partial\beta) = (\partial\Phi/\partial R) \tag{2.132}$$

## 2.21 Meridional Streamlines and Velocities

Consider the inlet of a radial turbine runner or centrifugal pump inlet as shown in Figure 2-25. The axial flow at the inlet plane that is parallel to the machine centerline axis and normal to a circle at the beginning of the flow contains all the streamlines passing through the machine. These streamlines form a surface that passes through the annular space between the shrouds.

At any point on the surface P, the flow velocity vector may be resolved into two components: a tangential component, $c_u$, and a through flow or meridional component, $c_m$. The tangential

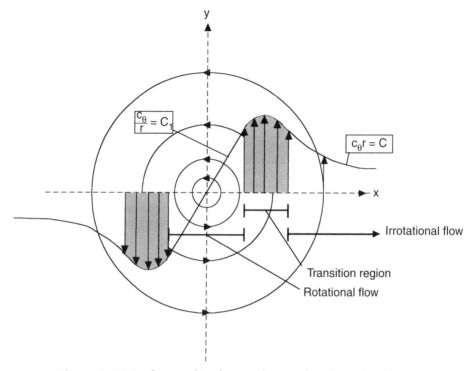

**Figure 2-23** Rankine combined vortex showing the velocity distribution.

component is in a plane normal to the axis, and the meridional component is that component in a longitudinal plane containing the axis. A cross section through which all the meridional streamlines pass normally is a flow cross section. A mathematical and graphical description of the three-dimensional meridional streamlines is somewhat difficult. In practice, a one-dimensional approximation is sufficient for most purposes.

## 2.22 Effects of Friction on Flows through Turbomachines

The assumption of frictionless flows through turbomachines is useful in a number of respects (e.g., visualization of flow patterns and identification of potential problem areas when designing turbomachines). However, the most important aspects of friction such as the development and interaction of boundary layers and separation of boundary layers cannot be completely analyzed theoretically. Some progress has been made with computer solutions by three-dimensional numerical modeling. A large number of software packages are available for 3-D modeling of turbulent flows around solid bodies and, although these provide better visual solutions of the flows and valuable data, models must still be designed and tested before a prototype machine can be built. Even then, empirical equations are still needed to predict such variables as scale-up and cavitation.

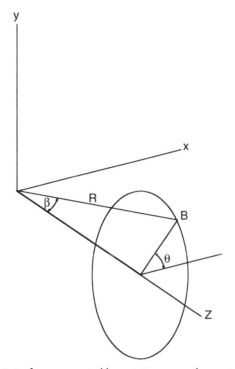

**Figure 2-24** Surface generated by rotation around an axis of symmetry.

This is not to invalidate the use of software numerical prediction, but in the end experimental data from prototype operation still remains the practical basis of other designs.

## 2.23 Solved Problems

### 2.23.1 Shape of the Surface of a Free Vortex

Calculate the shape of the surface h(r) of a free vortex (see Figure 2-26), given that the velocity potential is: $\Phi = U_0 r_0 \varphi = U_0 r_0 \arctan(y/x)$.

**Solution**
The velocity components are:

$$u_r = (\partial\Phi/\partial r) = 0 \tag{2.133}$$

$$u_\varphi = (1/r)(\partial\Phi/\partial\varphi) = U_0(r_0/r) \tag{2.134}$$

$$u_z = (\partial\Phi/\partial z) = 0 \tag{2.135}$$

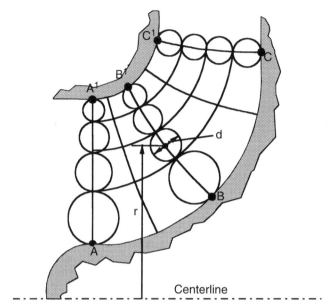

**Figure 2-25** Meridional streamlines for a one-dimensional flow.

The Bernoulli equation is applicable:

$$p + (\rho/2)u^2 + gz = \text{constant} \tag{2.136}$$

Applying Equation (2.136) between two points on the free surface, that is, as $r \rightarrow \infty$: $(z = 0 : p = p_0)$ and $(z = -h(r) : p = p_0)$, Equation (2.136) becomes:

$$p_0 + (\rho/2)u^2(r \rightarrow \infty) = p_0 + (\rho/2)\, u^2(r) - g\, h(r) \tag{2.137}$$

with $u^2(r \rightarrow \infty) = 0$

$$\therefore \quad h(r) = (U_0^2/2g)(r_0/r)^2 \tag{2.138}$$

**2.23.2** Velocity and pressure distribution in an inviscid, axisymmetric flow, given that the velocity distribution in an inviscid, axisymmetric plane flow is $u(r) = U_0(r/r_0)^n$

1. Determine the pressure distribution $p(r)$ when $p(r_0) = p_0$.
2. What is the value of $n$ such that the Bernoulli constant assumes the same value throughout the velocity field?

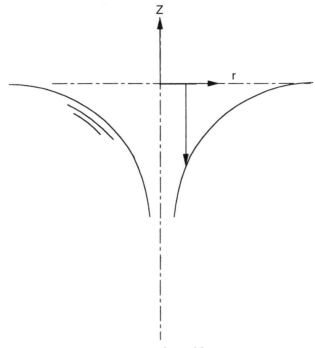

**Figure 2-26** Surface of free vortex.

## Solution

A sketch for the problem is given in Figure 2-27. The steady-state Euler equation without body forces may be written:

$$\rho u(\partial u/\partial \theta) = (\partial p/\partial \theta) \text{ in the direction of the pathline} \tag{2.139}$$

$$\rho(u^2/R) = (\partial p/\partial r) \text{ normal to the pathline} \tag{2.140}$$

Since the flow is symmetric $(\partial p/\partial \theta) = 0$. Therefore, Equation (2.139) may be written as an ordinary differential equation, that is,

$$\rho(u^2/R) = (dp/dr) \tag{2.141}$$

with $p(r_0) = p_0$

$$\therefore \quad p(r) - p(r_0) = \rho U_0^2/r_0^{2n} \int_{r_0}^{r} r^{2n-1} dr \tag{2.142}$$

When $n \neq 0$,

$$p(r) - p(r_0) = (\rho U_0^2/2n)[(r/r_0)^{2n} - 1] \tag{2.143}$$

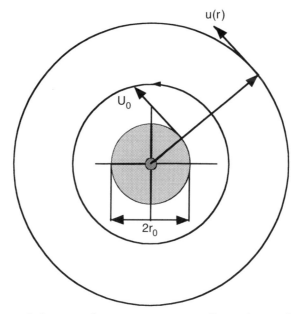

**Figure 2-27** Inviscid, axisymmetric rotating flow with central core.

When $n = 0$,

$$p(r) - p(r_0) = (\rho U_0^2) \ln(r/r_0) \tag{2.144}$$

For potential flow, the streamlines are concentric circles with $r = 0$ at their center:

$$p + (\rho/2)u^2 = \text{constant} = C \tag{2.145}$$

Consider two streamlines:

$$\text{at } r = r_0: \quad p(r_0) + (\rho/2)U_0^2 = \text{constant} = C_0 \tag{2.146}$$

$$\text{for } r > r_0: \quad p(r) + (\rho U_0^2)(r/r_0)^{2n} = C(r) \tag{2.147}$$

If the constant is the same for every streamline, then Equation (2.146) equals Equation (2.147).

$$\therefore \quad p(r) - p(r_0) = (\rho U_0^2)[(r/r_0)^{2n} - 1] \tag{2.148}$$

We see that n must equal $-1$ and therefore,

$$u(r) = U_0(r/r_0) \tag{2.149}$$

### 2.23.3 Potential Flow Past a Half-body

A three-dimensional, axisymmetric body is formed by combining a three-dimensional source of strength m = Q/4π with a three-dimensional uniform flow. The velocity and pressure at infinity are $V_\infty$ and $p_\infty$.

Determine:

1. The equation of the surface of the body
2. That part of the body on which the pressure is greater than $p_\infty$
3. The resultant force exerted by the fluid on the body, over the section where $p > p_\infty$

### Solution

The body formed by the combination of flows is shown in Figure 2-28.

Using spherical coordinates, we find that the velocity potential and stream function of a three-dimensional source are:

$$\Phi = -m/R \tag{2.150}$$

$$\Psi = -m\cos\beta \tag{2.151}$$

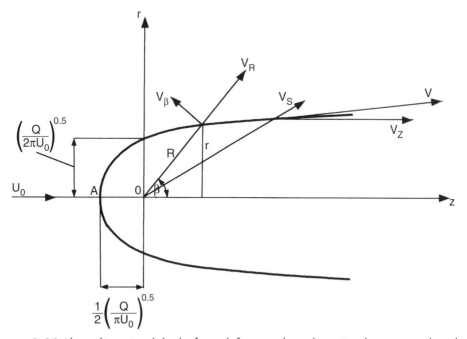

**Figure 2-28** Three-dimensional body formed from a three-dimensional source and a three-dimensional uniform flow field.

and for the three-dimensional uniform flow:

$$\Phi = U_0 \, R \cos \beta \qquad (2.152)$$

$$\Psi = (U_0/2) \, R^2 \sin^2 \beta \qquad (2.153)$$

By a suitable choice of origin, the constants normally associated with Equations (2.150) and (2.151) may be set to zero.

The stream function of the resultant combination of source and uniform stream is:

$$\Psi = -m \cos \beta + (U_0/2) \, R^2 \sin^2 \beta \qquad (2.154)$$

The velocity components of the flow are given by Equations (2.155) and (2.156):

$$V_R = m/R^2 + U_0 \cos \beta \qquad (2.155)$$

$$V_\beta = -U_0 \sin \beta \qquad (2.156)$$

A stagnation point exists where

$$V_R = V_\beta = 0 \qquad (2.157)$$

That is,

$$(m/R^2 + U_0 \, \cos \beta)^2 + (U_0 \sin \beta)^2 = 0 \qquad (2.158)$$

The only possible solution to Equation (2.159) is where $R = (m/U_0)^{1/2}$ and $\beta = \pi$, the point A, in Figure 2-28. Along the negative z-axis ($\beta = \pi$), Equation (2.154) becomes $\Psi_0 = +m$. Thus, the equation for the stream surface passing through the stagnation point is:

$$m = -m \cos \beta + (U_0/2) \, R^2 \sin^2 \beta \qquad (2.159)$$

Equation (2.159) may be rewritten as:

$$R^2 = (2m \, /U_0)[(1 + \cos \beta)/\sin^2 \beta] \qquad (2.160)$$

Noting that $r^2 = R^2 \sin^2 \beta$ Equation (2.160) may be written as:

$$r^2 = (2m \, /U_0) \, [(1 + \cos \beta)] \qquad (2.161)$$

This surface corresponds to the surface of a solid of revolution about the z-axis. We note that as $z \to \infty$, $\beta \to 0$. The asymptotic radius of the body at this point is: $r = r_0 = 2(m/ \, U_0)^{1/2}$.

The resulting body formed by this surface is called a half-body. The pressure distribution may be found from the Bernoulli equation:

$$p = p_\infty + \rho\, U_0^2/2 + \rho V^2/2 \tag{2.162}$$

The fluid velocity is the vector sum of $V_S$ and $U_0$ that is,

$$V^2 = V_S^2 + U_0^2 + 2U_0 V_S \cos\beta \tag{2.163}$$

$$V_S = (\partial\Phi/\partial R)_{source} = m/R^2 = U_0(r_0/2R)^2 \tag{2.164}$$

$$\therefore \qquad V^2 = U_0^2[(r_0/2R)^4 + 1 + 2(r_0/2R)^2 \cos\beta] \tag{2.165}$$

$$\text{Thus, } p = p_\infty - (\rho\, U_0^2/2)[(r_0/2R)^4 + 2(r_0/2R)^2 \cos\beta] \tag{2.166}$$

On the surface of the body, Equation (2.160) may also be written:

$$r^2 = (r_0^2/2)(1 + \cos\beta) = r_0^2 \cos^2(\beta/2) \tag{2.167}$$

$$\therefore \qquad (r_0/R) = [r/\cos(\beta/2)]/[r/\sin(\beta/2)] = 2\sin(\beta/2) \tag{2.168}$$

Equation (2.166) may be written with the aid of Equation (2.156) as:

$$p = p_\infty - (\rho U_0^2/2)2\sin^2(\beta/2)[(3/2)\sin^2(\beta/2) - 1] \tag{2.169}$$

$$p > p_\infty \text{ when } (3/2)\sin^2(\beta/2) > 1, \text{ that is, when}$$

$$\sin(\beta/2) > (2/3)^{1/2} \tag{2.170}$$

The resultant force acting on the boundary is in the direction of the stream. The force acting over a small area dA is:

$$F_z = \int (p - p_\infty)\, dA = \int (p - p_\infty)2\pi r\, dr \tag{2.171}$$

Substitution of $(p - p_\infty)$ from Equation (2.157) and rdr from Equation (2.149) yields:

$$F_z = \pi r_0^2 (\rho U_0^2/2) \int_0^{1/3} [3(r/r_0)^4 - 4(r/r_0)^2 + 1]\, d(r/r_0)^2 \tag{2.172}$$

Integrating:

$$F_z = (\pi r_0^2 \rho\, U_0^2/2)(4/27) \tag{2.173}$$

### 2.23.4 Homologous Reaction Turbines

A full-scale turbine is to be run at 300 rpm under a head of 60 m. A model one-sixth the size of the full-scale turbine is tested at 10 m head and develops 5 kW and with a volumetric flow rate of 0.06 m³/s. At what speed should the model be run, and what power should be obtained from the full-scale turbine assuming it is 5% more efficient than the model? What kind of turbine would be suitable for use under these operating conditions?

### Solution

For homologous turbines:

$$(H_1/H_2) = (U_1^2/U_2^2) = (D_1^2/N_1^2)/(D_2^2/N_2^2) \tag{2.174}$$

and

$$(N_1/P_1^{1/2})/(H_1^{5/4}) = N_S = (N_2/P_2^{1/2})/(H_2^{5/4}) \tag{2.175}$$

Substituting values in Equation (2.174) and solving for $N_1$: $N_1 = 735$ rpm

It is assumed that the model and prototype for this problem have the same maximum efficiencies. Substituting values in Equation (2.175): $P_2 = 2646\,\text{kW}$

Also:

$$P_1 = \gamma\, Q_1\, H_1\, \eta_1 \tag{2.176}$$

Substituting values, $\eta_1 = 0.85$ and $\eta_2 = 0.90$.
Therefore, the actual power developed $= 2646(0.90/0.85) = 2802$ kW.
Solving Equation (2.175) for $N_S$: $N_S = 95.0$.
This would correspond to the Francis turbine range.

### 2.23.5 Scaling of Hydraulic Turbines

A turbine in a river hydroelectric system is designed to produce 20 MW at a designed overall efficiency of 92% when it is running at 100 rpm. The effective head across the turbine is 20 m, and its runner OD is 3 m. A geometrically similar model has a runner of OD = 300 mm and is tested at 10 m head. What are the flow rate, the overall efficiency, and the power produced? Compare results by using equations that have different combinations of parameters:

1. The equation due to Hutton:

$$(1 - \eta_0)/(1 - \eta_{\text{model}}) = 0.3 + 0.7(\text{Re}_{\text{model}}/\text{Re}) \tag{2.177}$$

2. The equation due to Moody:

$$(1 - \eta_0)/(1 - \eta_{model}) = (H_{model}/H)^{0.01}(D_{model}/D)^{0.25} \qquad (2.178)$$

3. The equation due to Pfleiderer:

$$(1 - \eta_0)/(1 - \eta_{model}) = (Re_{model}/Re)^{0.01}(D_{model}/D)^{0.25} \qquad (2.179)$$

$$\eta_0 = \text{overall efficiency, and the subscript values refer to the model.}$$

$$Re = (\rho ND^2/\mu)$$

## Solution

The full-size flow rate is given by:

$$(20 \times 10^6)/0.92 = (20)(g)(10^3) Q \qquad (2.180)$$

$$Q = 110.8 \, m^3/s$$

Applying $Q/(ND^3) = $ constant and $gH/(N^3D^3) = $ constant and substituting the data:

$$N_{model} = 707 \text{ rpm and } Q_{model} = 0.78 \, m^3/s$$

Using Equation (2.177):

$$(1 - \eta_0)/(1 - \eta_{model}) = 0.3 + 0.7(100 \times 0.3^2/707 \times 3^2)^{0.2}$$

so that, $\eta_{model} = 83.6\%$
Using Equation (2.178):

$$(1 - \eta_0)/(1 - \eta_{model}) = (10/20)^{0.01}(0.3/3)^{0.25}$$

so that, $\eta_{model} = 85.7\%$
Using Equation (2.179):

$$(1 - \eta_0)/(1 - \eta_{model}) = (100 \times 0.3^2/707 \times 3^2)^{0.01}(0.3/3)^{0.25}$$

so that, $\eta_{model} = 84.8\%$
The average of these is 84.7%. Using this value, the model power developed, assuming the mechanical efficiencies remain the same, is:

$$P_{model} = (10)(g)(1000)(0.78)(0.848) = 64.9 \, kW$$

# 2.24 References

Anderson, H.H., Statistical records of pump and water turbine effectiveness. *Inst. Mech. Engrs. Conference, Sealing for Performance Prediction in Rotodynamic Machines*, September (1977).

Nixon, R.A., *Examination of the problem of pump scale Laws*, NEL Report, April (1965).

Nixon R.A., and Cairney, W.D., Scale effects in centrifugal cooling water pumps for thermal power stations, NEL Report 505, April (1972).

*C H A P T E R   3*

# TURBINES

## 3.1 Classification of Turbines

Specific speed of turbines and pumps has been defined in Chapter 2, and there is no doubt that specific speed is the most useful way of classifying turbines. Such a classification may be summarized as:

1. Impulse turbines of low specific speed $0 < N_S < 12$. High head, low-flow rate.
2. Radial-flow turbines having Francis-type runners in the approximate range, $N_S = 20–100$. Medium head, medium-high flow rate.
3. Axial-flow, propeller/Kaplan-type runners having $N_S > 100$. Low head, low-medium flow rate.

$N_S$ in the above classification has units of N—rpm, P—bhp, H—ft.

For units in the SI system, that is, N—rpm, P—kW, H—m, the above values must be multiplied by 3.812. A classification based on the above ranges of specific speed is useful in that a preliminary decision may be made as to which turbine is most suitable for a given application (see Figure 2-4).

Figure 3-1 is also useful in this regard. Here specific speeds, $N_S$, of different turbines are shown plotted as a function of head across the machine. They are given as area plots. Notice that the scale is log-log. Although the bottom area, for Pelton wheels, is largest, it has the narrowest range of applicability. The other two areas for Francis and Kaplan turbines have a much greater range of applicability. The numerous other turbines that have been invented lie within this broad range and in some cases overlap the ranges of the three main turbines.

All rotary pumps can be operated in reverse to act as turbines, and all rotary turbines can be operated in reverse to act as pumps. The foregoing statement is not true of all turbines. Impulse turbines, for example, are jet reaction turbines, and it is not possible to reverse the flow.

## 3.2 General Operating Conditions

The best practical operating condition of a turbine is usually at constant speed, with a gate opening (controlling flow) that is dependent on the head across the machine. Variation of static head is

**55**

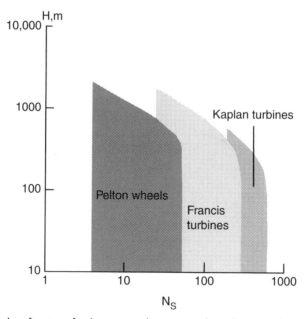

**Figure 3-1** Broad classification of turbines. Head, H, across the turbine as a function of specific speed $N_S$ (Units: rpm, kW, m)—log-log plot.

particularly important in low head plants where the tailwater level may rise in time of flood at a different rate to the headwater level, thus giving a significant decrease in power. Figure 3-2 shows typical efficiency curves at constant speed as a function of rated power for constant rotational speed and constant head.

## 3.3 Impulse Turbines-Pelton Wheels

An impulse turbine is a turbomachine in which kinetic energy from one or more fast-moving jets is converted to rotational mechanical energy delivered to the shaft of the machine. Several types of impulse turbines have been invented, but only one has survived in appreciable numbers to the present day. This is the Pelton wheel. A typical Pelton wheel system is shown schematically in Figure 3-3.

The high-speed jets of fluid impinge, in as shock-free a way as possible, on vanes or buckets located around the periphery of a wheel. Figure 3-4 shows a jet and the velocities associated with it. The energy delivered by the jet to the buckets is dissipated as head losses in several ways:

1. Nozzle head loss, given by:

$$(1/C_V^2 - 1)(V_1^2/2g) \tag{3.1}$$

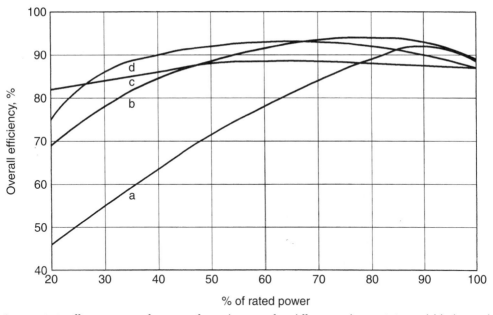

**Figure 3-2** Efficiency as a function of rated power for different turbines. (a) Fixed blade axial; (b) Francis; (c) Impulse; (d) Kaplan.

2. Fluid friction—friction head lost on the surface of the buckets:

$$h_f = kw_1^2/2g \qquad (3.2)$$

3. Lost kinetic energy:

$$h_{KE} = V_2^2/2g \qquad (3.3)$$

Therefore, the theoretical head delivered to the buckets h is:

$$h = H - (1/C_V^2 - 1)(V_1^2/2g) - kw_1^2/2g - V_2^2/2g \qquad (3.4)$$

There is also windage loss—drag losses of the water/air on the rotating wheel and mechanical friction loss.

### 3.3.1 Speed Factor, Φ

It can easily be shown that with no losses the maximum power that can be obtained from a Pelton wheel occurs when the bucket velocity is equal to half the jet speed. Wheel losses are usually

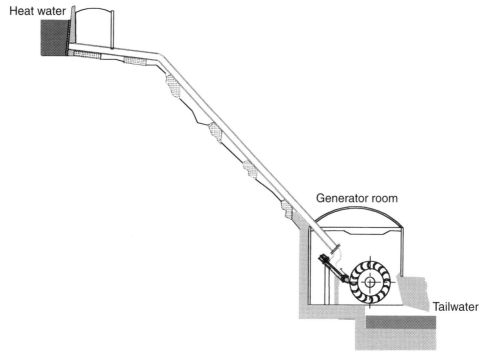

**Figure 3-3** A typical Pelton wheel set-up, showing the high head and relative locations of the reservoir and generating station.

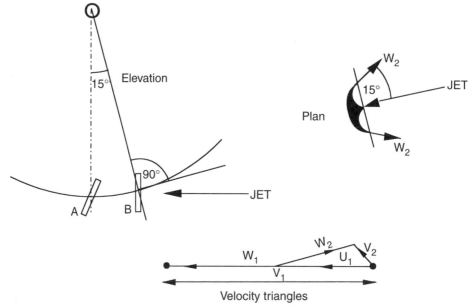

**Figure 3-4** Pelton wheel bucket showing jet action and relative velocities and typical values of angles.

expressed as:

$$\Phi = U/(2gH)^2 \qquad (3.5)$$

Because U is a function of N, the rotational speed, and N is part of the equation defining $N_S$, it follows that $\Phi$ must be a function of $N_S$. A cross section of a Pelton wheel bucket is symmetric; this allows for symmetry of the impinging jet and thus even thrust on the bucket.

### 3.3.2 Specific Speed of Pelton Wheels

The specific speed of a turbine (Equation 2.37) was defined as:

$$N_S = NP^{0.5}/H^{5/4} \qquad (3.6)$$

The rotational speed (rpm) is:

$$N = (60u)/(\pi D) = 60\,\Phi(2gH)^{0.5}/(\pi D) \qquad (3.7)$$

The volumetric flow rate is:

$$Q = V_1(\pi d^2/4) \qquad (3.8)$$

$V_1$ = jet velocity and d = nozzle diameter

Power developed is:

$$P = \gamma QH\eta_o \qquad (3.9)$$

Combining Equations (3.6) to (3.9) and rearranging, we obtain:

$$N_S = K(D/d) \qquad (3.10)$$

K in Equation (3.10) is a combination of all the variables, but since they vary only slightly, K varies slightly, so that the specific speed depends primarily on (D/d). If (D/d) is relatively large, the cost of the wheel becomes proportionately greater; bearing friction is increased together with windage loss. On the other hand, if (D/d) becomes small, the bucket dimensions become unreasonable in terms of the wheel diameter. However, before this point is reached, there will be an increased departure from tangential action of the jet and the efficiency will fall. An equation due to Daily (1950) for optimal values of the parameters of Equation (3.9) is:

$$(D/d) = (206/N_S) \qquad (3.11)$$

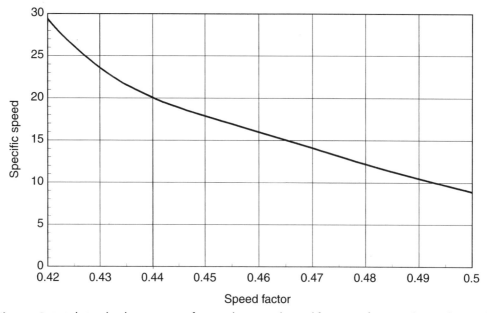

**Figure 3-5** Relationship between specific speed, $N_S$, and speed factor, $\Phi$, for a single nozzle wheel. Test data from Quick (1940).

Equation (3.11) is in good agreement with the experimental data of Quick (1940). Figure 3-5 shows the relationship between $\Phi$ and $N_S$ for a single nozzle wheel. For multiple nozzles, the value of $N_S$ should be multiplied by the square root of the nozzle number.

### 3.3.3 Nozzles

Altering the flow through the nozzle by changing the effective jet area regulates the power output of a Pelton wheel. This is done by a bulb or spear that can move to and fro along the axis of the nozzle together with deflection of the jet. Figure 3-6 illustrates this sort of arrangement. By changing the effective jet area, the jet velocity and the nozzle efficiency are changed simultaneously. Figure 3-7 shows the effect of jet velocity change in terms of change in bucket/jet velocity ratio. This in turn affects the overall efficiency of the wheel. Figure 3-8 shows in a general way that the nozzle efficiency is affected by jet area change. The exact shape of the curve is a function of the nozzle design.

### 3.3.4 Jet Force on Runner

Because the buckets rotating around a central axis continuously enter and leave a jet, the net force on a bucket rises and falls from zero to a maximum and then back to zero. The maximum value is

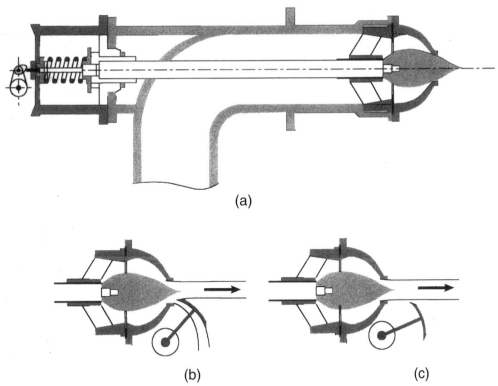

(a)

(b)                                    (c)

**Figure 3-6** (a) Typical nozzle for a Pelton wheel, needle closed; (b) needle open with deflector in a position that is deflecting part of the jet; (c) needle open with deflector not touching the jet.

somewhat flat over a small angular range; this is when the plane of the bucket is approximately normal to the axis of the jet. This is illustrated in Figure 3-9 as the solid curve B. Neighboring dotted curves are also shown as A and C. The net effective force on the runner is the top curve of Figure 3-9.

### 3.3.5 Arrangement of Nozzles and Size of Jets

When the generator shaft is horizontal, the simplest machine would have one jet and the runner overhung on the generator shaft. In most cases, however, the power required to be generated is too large for a single nozzle. In addition, the rotational speed would have to be reduced to an unacceptable level. The alternatives are to have two runners on the same shaft or multiple nozzles, up to six in number for one runner. The nozzles are spaced at equal angular increments around the runner.

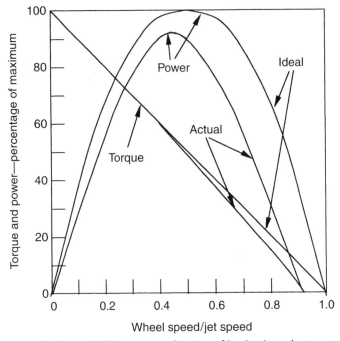

**Figure 3-7** Wheel overall efficiency as a function of bucket/jet velocity ratio (U/V$_J$).

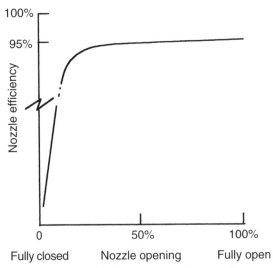

**Figure 3-8** Nozzle efficiency as a function of jet area.

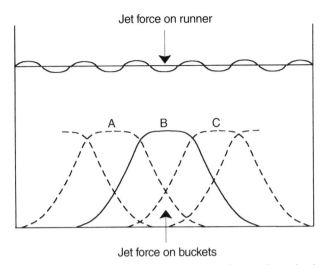

Jet force on runner

Jet force on buckets

**Figure 3-9** Forces on runner and buckets for a Pelton wheel.

When large power is to be generated by a single generator, the generator shaft is usually horizontal and connected to two single jet turbines.

### 3.3.6 Jet Velocity and Diameter

The required volumetric flow rate for power output of P kW is given by:

$$Q = (1000\,P)/(\eta\gamma H) \tag{3.12}$$

where: $\eta$ = overall efficiency of the turbine, usually 0.88 to 0.90.

The Q given by Equation (3.12) gives the total jet area. The jet velocity is given by:

$$V_J = C_V(2gH)^{1/2} \tag{3.13}$$

$C_V$ usually lies in the range 0.97–0.98. If the number of nozzles is n, the volumetric flow rate through each nozzle is:

$$q = Q/n \tag{3.14}$$

$$q = [(\pi/4)d^2](V_j) \tag{3.15}$$

It is usual to express the nozzle flow as a unit flow rate:

$$q^1 = q/H^{1/2} \tag{3.16}$$

Thus, the jet diameter is given by:

$$d = K(q^1)^{1/2} \tag{3.17}$$

The constant K in Equation (3.17) is evaluated from Equations (3.12) to (3.16).

The part-load characteristics of Pelton wheels are good; often this is one of the primary factors that influence the choice of machine. Alternator speed and total head across the machine fixes the ratio $(U/V_J)$. Power is regulated by the volumetric flow rate, which in turn is controlled by the effective nozzle area. Desirable characteristics such as high part-load efficiency and a flat regulation curve ensue. On occasion, the load on the machine will be suddenly reduced on a machine. The effect is to increase the rotational speed equally rapidly. Rapid valve closure to control this is impractical because of water hammer. Surge tanks are impractical, usually because of the normal high heads across the machine.

Three methods are in common use:

1. Deflector plates, either straight or curved, which decrease the flow rapidly and which may be easily controlled by a speed regulation governor.
2. Main jet deflection by means of an auxiliary jet.
3. Disintegration of the jet by imparting a swirl to the flow prior to nozzle entry, thus increasing the jet area.

### 3.3.7 Runner

The pitch circle of the runner is the circle to which the axis of the jet is tangential. The central axis of the jet, nozzle spear, and nozzle itself must be aligned in such a way that it intersects the central plane of the bucket. The nozzle should be at a distance from the axis of rotation, which is half the pitch circle diameter. The velocity of the runner at its pitch circle diameter is given by Equation (3.5): that is, $\Phi = U/(2gH)^2$. A good design value for $\Phi$ is 0.46–0.47. Thus, the runner diameter may be calculated in two ways: (1) by using the assumed design $\Phi$ and the equation:

$$D = (60u)/(\pi N) \tag{3.18}$$

or (2) by using Equation (3.11).

The power required at a given head means that the operating speed of the turbine should be as high as the turbine and generator will allow. The components must have sufficient strength with an adequate factor of safety, not only at the normal operating speed, but also for the runaway speed which is usually 90% higher. It has been found that for a large number of tests on impulse turbines the maximum power is generated in the range 300 to 750 rpm (Gray, 1958).

An example of a Pelton turbine with six injectors is shown in Figure 3-10. An auxiliary jet acting as an injector brake has been fitted to this system, as shown in the plan view in Figure 3-11.

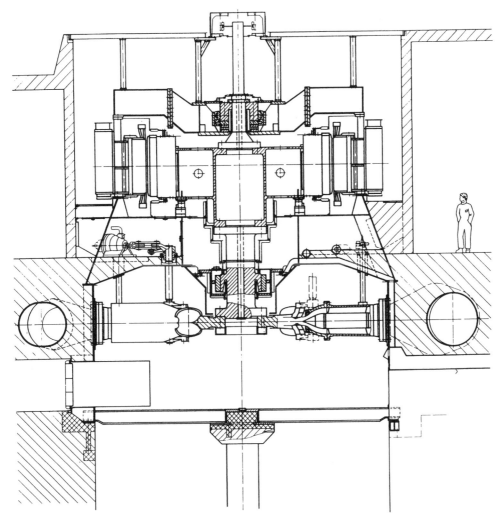

**Figure 3-10** Side view of one of eight Pelton turbines installed in San Carlos, Colombia. Designed by Sulzer-Escher Wyss SA. Zurich. Nominal power: 170.4 MW. Nominal head: 578 m. Nominal Q = 33.37 m$^3$/s. (Diagram courtesy VA TECH HYDRO)

## 3.3.8 Turgo Wheels

The Turgo wheel is a unique impulse turbine somewhat similar to the Pelton wheel. The basic difference is the construction of the runner, as illustrated in Figure 3-12. The runner is cast in one piece with single-discharge buckets held inwardly by the hub and outwardly by a flat band on the periphery. The jet impingement and path in Figure 3-12(b) is different from that of the Pelton

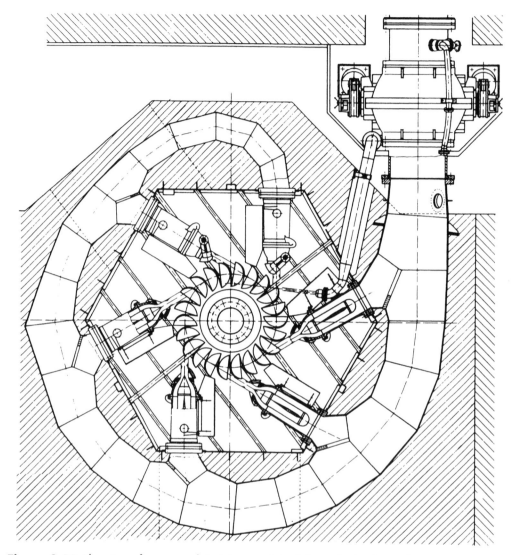

**Figure 3-11** Plan view of runner and six injectors of the Pelton wheel turbine of Figure 3-10. Note auxiliary jet. (Diagram courtesy VA TECH HYDRO)

wheel in Figure 3-12(a). In effect, the runner is a modified Francis runner with jet feed from a Pelton-type nozzle. The number of jets is usually one and occasionally two.

Because of the hybrid nature of the Turgo wheel, its specific speed would be expected to be higher than a typical Pelton wheel and approaching the lower range of a Francis turbine. $N_S$ is in the range 10 to 125/130. The Turgo wheel is therefore a useful machine that bridges the gap between Pelton wheels and Francis turbines. Output power up to 3000 kW has been obtained.

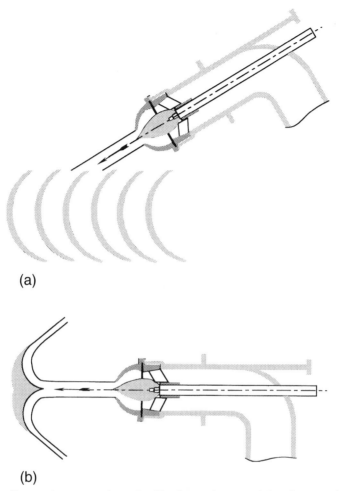

(a)

(b)

**Figure 3-12** Difference between Pelton wheel bucket and injector (a) and Turgo wheel blading and injector (b).

## 3.4 Radial-Flow Turbines—Francis Turbines

The inward-flow radial turbines in use from the nineteenth century to the present and virtually the same as present-day turbines were developed by J. B. Francis and are commonly known as Francis turbines. Schematic views of a typical Francis turbine are shown as a plan view in Figure 3-13 and as a side elevation in Figure 3-14.

A two-dimensional cut through inlet and outlet shows the appropriate velocity triangles; this may be seen in Problem 3.8.4. The figure with this problem shows typical values of an inward-flow Francis turbine. When we consider the Euler turbine equation, one part of the equation that affects the power generated is the second term, $c_{U2}\, U_2$. If $c_{U2}\, U_2$ could be made zero—that is,

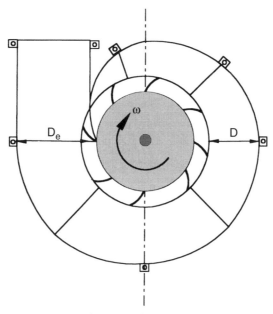

**Figure 3-13** Plan view of a typical Francis layout.

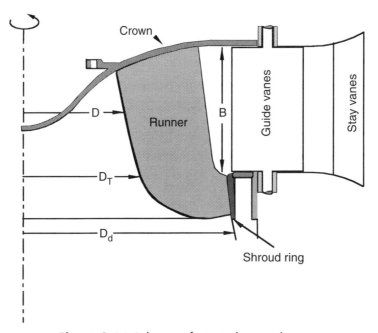

**Figure 3-14** Side view of a typical Francis layout.

if we try to make $c_{U2} = 0$—this will have the desired effect of increasing the power output. It is not usually possible to do so, but a reduction in its value close to zero is desirable.

The flow through a Francis turbine is three dimensional. The flow prior to entry into the guide vanes is a swirling (vortex) flow, with increasing average velocity as the fluid goes around the scroll tube because of area decrease. Because the head and flow rate can vary from day to day, the velocity triangle at input to the runner must be adjusted to optimize power output. This is done by adjusting all the guide-vane angles simultaneously to accommodate head and flow rate changes. Data from sensors for head and flow are fed to servomotors that control the blades. A mechanism for doing this is illustrated in Figure 3-15.

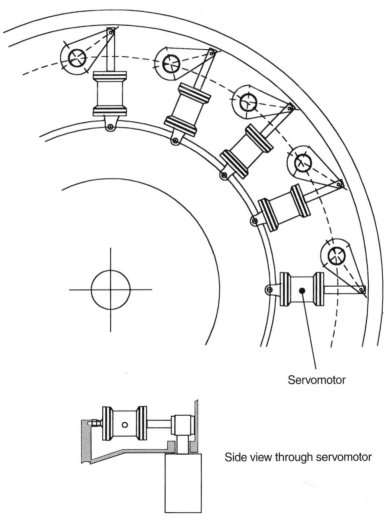

Servomotor

Side view through servomotor

**Figure 3-15** Plan and side views of guide-vane adjustment mechanism.

For equal power and head, Francis turbines require a smaller installation space than impulse turbines. Also, the highest efficiency is obtained at near full load. This compares with impulse turbines that have their best efficiencies at near half load, with the efficiency falling off gradually as full load is approached. Thus, a Francis turbine is more efficient where most energy is generated, but it is less efficient at small loads.

Francis turbines may be arranged with the shaft either vertical or horizontal. The vertical shaft is preferable to the horizontal because it permits the turbine to be placed at a lower level relative to the tailwater level. In addition, from a manufacturing point of view, it is more economic to have a vertical shaft for a large machine.

The heart of a Francis turbine is the runner. So the design of the machine and its ancillary parts is focused on this to obtain maximum efficiency. The maximum efficiency is reached when all losses are kept to a minimum.

Losses incurred are:

1. Vortex formation at the runner exit
2. Seal leakage
3. Friction losses in bearings and glands
4. Friction losses in the guide vanes, casing, and draft tube.

### 3.4.1 Choice of Turbine Speed

Once the required output of the turbine has been chosen, the highest possible speed is obtained from:

$$N = N_S H^{5/4} / P^{1/2} \tag{3.19}$$

where $N_S$ is the highest specific speed for H. The nearest lowest synchronous speed for the generator becomes the highest practical speed for the set $N_{PRAC}$. From this the practical value of $N_S$ is calculated:

$$N_S = N_{PRAC} P^{1/2} / H^{5/4} \tag{3.20}$$

### 3.4.2 Effect of Gate Opening

Draft tubes are dealt with in detail in Chapter 8. However, important effects that occur in the draft tube are initiated at the gate. At partial guide-vane opening, the tangential velocity component, $c_{U2}$, is large, resulting in a relatively low pressure at the runner inlet. This results in a dead water space following the runner inlet (illustrated in Figure 3-16 as region 2). The main flow occurs in region 1. At the interface of the dead water and the moving liquid, a train of vortices occurs; this is initiated slightly downstream of the runner. At a larger guide-vane opening, but still less than 50% of the full opening, the dead water space is reduced and the vortex train consolidates into a single larger vortex, indicated by the dotted lines in the vortex core in Figure 3-16.

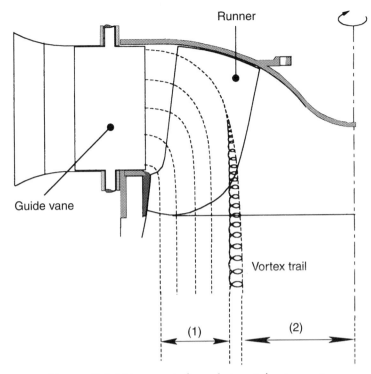

**Figure 3-16** Vortex train formed at partial gate opening.

The larger core vortex is unstable and oscillates about a point of suspension on the inner top wall of the turbine case rather like a suspended swinging rope. The vortex core also shows precession. This periodic movement of the core gives rise to cyclical pressure changes occurring roughly once in every six runner revolutions. These pressure changes manifest themselves as pulsations in the power output of the generators.

As a remedy for this effect, air may be introduced into the draft tube through an automatic air valve at part guide-vane opening. The air fills the space in Figure 3-17, and because of expansion at low pressure the flow is similar to that of Figure 3-16. Losses due to the large oscillating vortex core are averted.

Two commercial installations of Francis turbines are shown in Figure 3-18 and Figure 3-19.

## 3.5 Axial-flow Turbines—Propeller and Kaplan Turbines

The propeller turbine design was originally motivated by the need to develop high specific speed machines for use in relatively low head situations where it would be uneconomic to use a Francis turbine. Much experimentation soon showed that the efficiency curves for propeller turbines were markedly peaked and therefore had a limited $N_S$ range. Following a number of model experiments

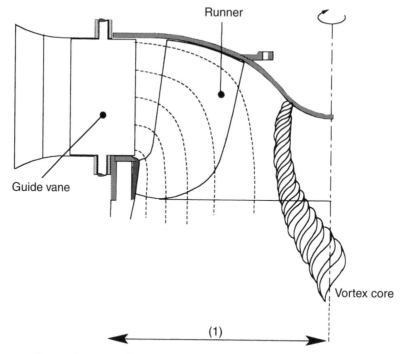

**Figure 3-17** Oscillating vortex core formed at partial gate opening.

Viktor Kaplan (1876–1934), an Austrian engineer, realized that changing the pitch of the blades could make a turbine with a greater range of applicability. In 1913, Kaplan designed a variable-pitch propeller turbine, the Kaplan turbine. Since that time, the operating head of the Kaplan turbine has been increased, and smaller Kaplan turbines have been used for heads as high as 65 m. The Kaplan turbine runner is hydraulically similar to the propeller turbine runner except that the hub is larger to accommodate the mechanism for blade angle shifting. The servomotor to accomplish this is located in the hub in some designs.

Three designs for mechanisms exist for blade operation:

1.  One design from 1922 has the servomotor located in the hub above the axes of the blades. Reciprocating motion of the servomotor piston is transferred to a trunnion mounted on a ring bolted to each blade.
2.  Another type has the servomotor located in a bulge in the shaft remote from the runner. The servomotor piston is connected by a rod to a crosshead located in the hub below the turning axis of the runner blades. Each blade has a lever keyed to its shaft with a connecting link to the crosshead.
3.  A third type, introduced in 1940, has the servomotor located in the nosepiece below the runner. The mounting is similar to 2 above.

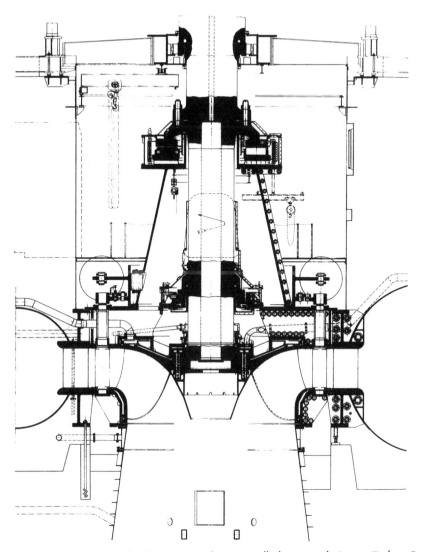

**Figure 3-18** Side view of one of eight Francis turbines installed at Ataturk Center, Turkey. Designed by Sulzer-Escher Wyss SA. Zurich. Nominal power: 306 MW. Nominal head: 151.2 m. Nominal Q = 218.5 m$^3$/s. (Diagram courtesy VA TECH HYDRO)

### 3.5.1 Combinator

The mechanism for controlling the relationship between the guide-vane angle and the runner blade angle is called a blade-control valve or Combinator. Mechanical connections from the piston of the runner-blade servomotor or the governor are taken to this valve. A cam connects with a linkage

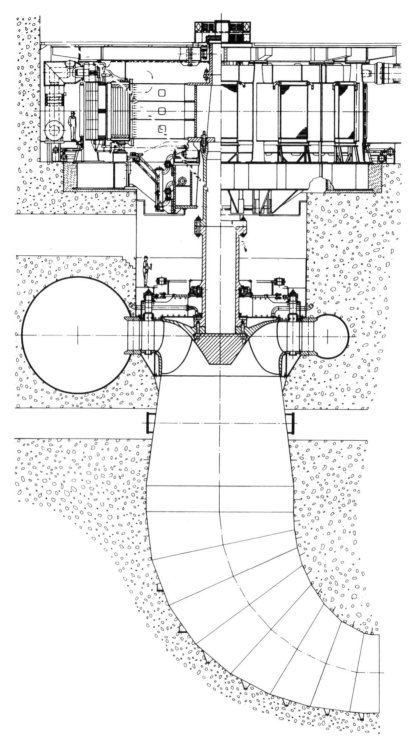

**Figure 3-19** Side view of one of four Francis turbines installed in Abbottabad, Pakistan. Designed by Sulzer-Escher Wyss SA. Zurich. Nominal power: 440 MW. Nominal head: 117.4 m. Nominal Q = 428.4 m³/s. (Diagram courtesy VA TECH HYDRO)

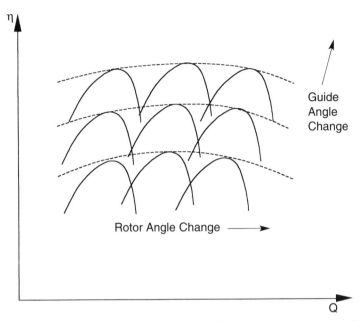

**Figure 3-20** Effects of rotor and guide-vane angle adjustments on turbine efficiency.

controlling oil pressure in the valve. The cam is shaped so that the correct relation between guide-vane angle and blade angle is obtained; valve oil pressure correctly positions the cam.

### 3.5.2 Effects of Rotor and Guide-vane Angle

The ability to change both runner and guide-vane angles simultaneously and in unison enables the operator of the turbine to achieve maximum efficiency under varying flow conditions. The effects of such changes for a Kaplan turbine are illustrated in Figure 3-20. There is an optimal curve for each of the angle adjustments and therefore a global optimum for the combination.

### 3.5.3 Selection of Speed and Runner Dimensions

The decision to install a Kaplan turbine must take into immediate consideration the hydraulic characteristics of the system, together with a full understanding of the requirements. It may well be that some of the hydraulic characteristics are in conflict. For example, if a full output is required over a wide range of head, the dimensions of the turbine will increase as the head variation increases. If the headwater level is fairly constant and net head variations are no more than ±10%, then close to full output can be obtained at the design point.

As a guide to preliminary design, four curves are presented for Kaplan and propeller turbines: Figures 3-21 and 3-22 are for Kaplan turbines, and Figures 3-23 and 3-24 are for propeller

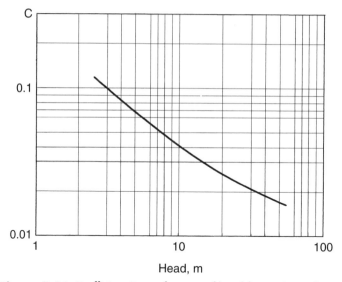

**Figure 3-21** Coefficient C as a function of head for Kaplan turbines.

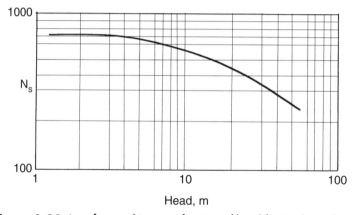

**Figure 3-22** Specific speed $N_S$ as a function of head for Kaplan turbines.

turbines. These represent the averages of a wide range of experimental data. The values of C in Figures 3-21 and 3-22 are calculated from Equation (3.21).

The variable C in these plots is given by:

$$C = D/(P)^{1/2} \qquad\qquad (3.21)$$

where:

P = power in kW
D = runner diameter in m

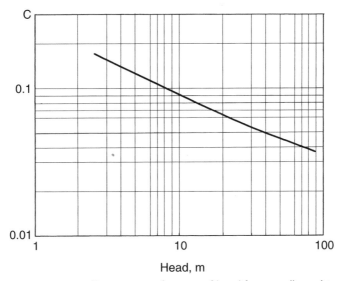

**Figure 3-23** Coefficient C as a function of head for propeller turbines.

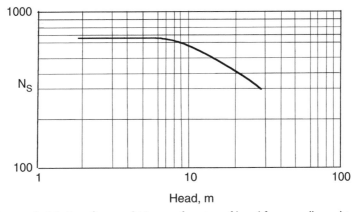

**Figure 3-24** Specific speed $N_S$ as a function of head for propeller turbines.

These plots are based on data obtained from operating commercial turbines. The variation of specific speed between different designs is considerable; the variation for runner diameter is much smaller. Therefore, an estimation of turbine dimensions based on a runner correlation rather than a specific speed correlation should be more accurate. A typical commercial Kaplan installation is shown in Figure 3-25.

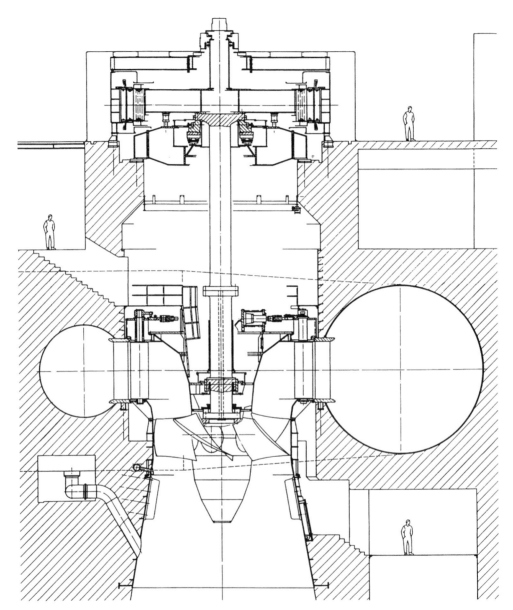

**Figure 3-25** Side view of one of six Kaplan turbines installed in Kwara, Nigeria. Designed by Sulzer-Escher Wyss SA. Zurich. Nominal power: 96.35 MW. Nominal head: 27.65 m. Nominal Q = 376.5 m$^3$/s. (Diagram courtesy VA TECH HYDRO)

# 3.6 Other Turbines

### 3.6.1 Pump Turbines (see Figures 3-26, 3-27, and 3-28)

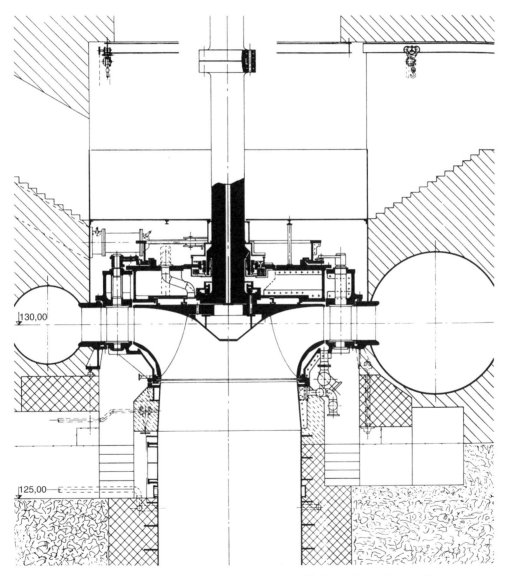

130,00

125,00

**Figure 3-26** Side view of one of three pump turbines installed 200 km northeast of Beijing, China. Designed by Sulzer-Escher Wyss SA. Zurich. (Diagram courtesy VA TECH HYDRO)

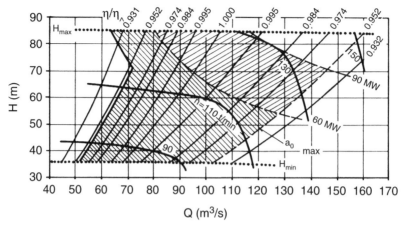

**Figure 3-27** Characteristics of a pump in one of the three pump turbines of Figure 3-26. Designed by Sulzer-Escher Wyss SA. Zurich. (Diagram courtesy VA TECH HYDRO)

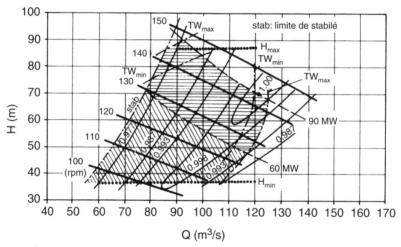

**Figure 3-28** Characteristics of a turbine in one of the three pump-turbines of Figure 3-26. Designed by Sulzer-Escher Wyss SA. Zurich. (Diagram courtesy VA TECH HYDRO)

### 3.6.2 Deriaz Turbine

The Deriaz turbine illustrated in Figure 3-29 is a mixed-flow radial turbine with adjustable runner blades. It is a hybrid machine possessing the characteristics of a Francis and a Kaplan turbine. One of its advantages is its flat characteristic over a wide range of partial load conditions.

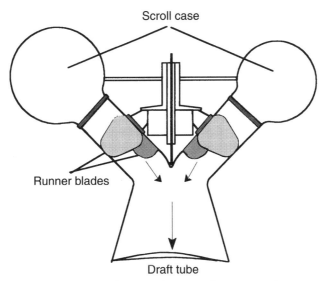

Figure 3-29 Schematic diagram of a Deriaz turbine.

### 3.6.3 **Bulb Turbine** (see Figure 3-30)

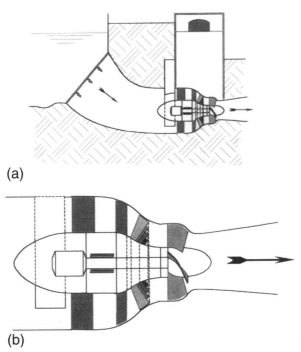

(a)

(b)

**Figure 3-30** Schematic diagram of a bulb turbine. (a) Shows installation (b) magnified view of turbine.

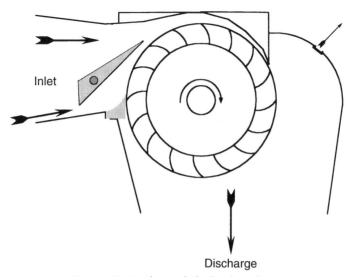

Inlet

Discharge

**Figure 3-31** The Banki hydraulic turbine.

### 3.6.4 Banki Turbine

The Banki turbine, named after its inventor, Donat Banki, is a radial-flow wheel and is a variant of a famous undershot waterwheel designed and built by Jean Victor Poncelet. The Poncelet wheel, a vertical one, had an angled sluice generating a broad jet of fluid that intersected the blades of the wheel in a tangential manner. It may be regarded as an early pressureless turbine. Poncelet's results of his theoretical and experimental studies were published in 1825, for which he was awarded the Mechanical Prize of the French Royal Academy of Sciences. Thereafter many Poncelet turbines were built and operated successfully.

The Banki turbine has a characteristic speed range between that of a Pelton wheel and a Francis turbine. Mockmore and Merryfield (1949) have made a thorough investigation of its characteristics. The turbine, schematically shown in Figure 3-31, consists of a wheel, like Poncelet's, with curved blades around the periphery. One of the differences is the mechanism of flow control.

### 3.6.5 Michell Turbine

The Michell turbine may be supplied by an open flume or pipe and was designed to replace old waterwheels. The similarity between it and the Banki turbine is evident. (See Figures 3-32 and 3-33 for an illustration of the Michell turbine.)

The Michell turbine has a horizontal shaft runner positioned close to the tailwater and is suitable for low heads. An automatic governor, as shown in Figure 3-32, regulates the discharge.

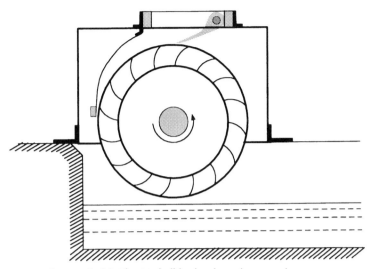

**Figure 3-32** The Michell hydraulic turbine—side view.

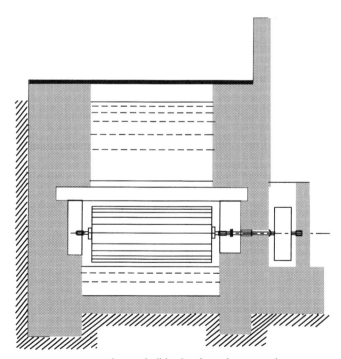

**Figure 3-33** The Michell hydraulic turbine—end view.

# 3.7 Control and Governing of Turbines

## 3.7.1 Function of a Governor

Power-demand changes on a turbine from the grid change the speed of the generating plant. A sudden rise in demand causes a drop in speed and vice versa. The correction for these demands is made through various mechanisms connected to guide vanes in the case of radial-flow and axial-flow turbines and to the spear and deflector position in the case of impulse turbines. These event corrections do not, of course, occur instantaneously; there is a time delay between the initial demand change from the grid and the necessary corrections to be made. For example, a decrease in demand from the grid is essentially instantaneous; the turbine correction decreases more slowly because of governor time lag. The speed change would be very considerable if it were not for the restrictions imposed by the flywheel effects of the rotating masses.

A further difficulty with governing is associated with the creation of strong pressure waves or water hammer by the rapid opening or closing of valves. This effect is exactly the opposite of what is required, causing the turbine to vary in the opposite direction to what is needed. This makes governing difficult.

## 3.7.2 Equations for Load Changes

The total kinetic energy possessed by the turbine set, that is, turbine, speed increaser, and generator, is given by:

$$E = \frac{1}{2} I \omega^2 \qquad (3.22)$$

where:

$I$ = moment of inertia of the rotating masses = $MR^2$
$M$ = mass
$R$ = radius of gyration
$\omega$ = angular rate of rotation = $2\pi N/60$

At steady-state conditions, we may write:

$$\eta_t P_t = P_G = P_0 \qquad (3.23)$$

where:

$\eta_t$ = turbine efficiency at time t
$P_t$ = input power to the turbine
$P_G$ = power demand by the grid
$P_0$ = power supplied by the turbine before load change

If $\eta_t P_t$ is $>$ or $<$ $P_G$ then a certain amount of energy is transferred to or from the turbine in time, t, by an amount $\Delta E$, which is a fraction of $E_t$, the value of E in Equation (3.22) at time t.

∴ We may write:

$$\Delta E = \int_0^T (\eta_t P_t - P_G)\, dt \qquad (3.24)$$

If we write indices 0 and 1 for the time at which the load changed from the instant of load change (time $= -0$) and 1 for the instant following the load change, then:

$$\Delta E = E_1 - E_0 = \frac{1}{2} I(\omega_1^2 - \omega_0^2) \qquad (3.25)$$

Thus:

$$\frac{1}{2} I(\omega_1^2 - \omega_0^2) = \int_0^T (\eta_t P_t - P_G)\, dt \qquad (3.26)$$

Or:

$$\frac{1}{2} MR^2 (2\pi/60)^2 (N_1^2 - N_0^2) = \int_0^T (\eta_t P_t - P_G)\, dt = K\,\Delta P_0 T \qquad (3.27)$$

where:

$$K = \left[ \int_0^T (\eta_t P_t - P_G)\, dt \right] \Big/ (\Delta P_0 T) \qquad (3.28)$$

$$(N_1^2/N_0^2) - 1 = K(\Delta P_0 T) \Big/ \left[ \frac{1}{2} MR^2 (2\pi/60)^2 N_0^2 \right] \qquad (3.29)$$

If we write $N_1 = N_0 \,\forall\, \Delta N$, then:

$$(N_0 \,\forall\, \Delta N)/N_0 = \left\{ 1 + \left\{ K(\Delta P_0 T) \Big/ \left[ \frac{1}{2} MR^2 (2\pi/60)^2 N_0^2 \right] \right\} \right\}^{1/2} \qquad (3.30)$$

For load rejection:

$$\Delta N/N_0 = \left\{ 1 + \left\{ K(\Delta P_0 T) \Big/ \left[ \frac{1}{2} MR^2 (2\pi/60)^2 N_0^2 \right] \right\} \right\}^{1/2} - 1 \qquad (3.31)$$

For load increase:

$$\Delta N/N_0 = 1 - \left\{ 1 + \left\{ K(\Delta P_0 T) \Big/ \left[ \frac{1}{2} MR^2 (2\pi/60)^2 N_0^2 \right] \right\} \right\}^{1/2} \qquad (3.32)$$

### 3.7.3 Governors

Single machines supplying power to small AC systems where accurate frequency control is necessary require a governor regardless of the load. The governor operates purely as a mechanism for maintaining constant speed regardless of the load. Thus, if the speed changes in response to a load change, then the flow rate is adjusted to suit the new load. The governing of the machine under these conditions is said to be astatic. Machines connected in parallel cannot be governed in this way because the load is not properly shared. If generator frequency is to be maintained constant, the speed of the output shafts changes. This is done by use of a mechanism known as a *speeder gear*. Turbines using this mechanism are said to be output-controlled. Large turbines have additional problems. Because of the large inertia forces inherent in such turbines, servomotors operated by oil pressure are needed. The piston motion of the servomotor is connected to the guide vanes of the turbine; an oil-pressure regulating valve, which in turn is connected to an actuator-connected pendulum damping gear and speeder gear, controls the piston motion of the servomotor. Rapid accelerations and decelerations resulting in rapid speed changes are thus kept to a minimum.

### 3.7.4 Relief Valves

Relief valves have the primary purpose of diverting a portion of the discharge so that the guide can be closed quickly without incurring unacceptable changes in speed and pressure. Without a relief valve, in a high-head plant the flywheel effect afforded by the generator would be excessive. Figures 3-34 and 3-35 show two designs. Figure 3-34 shows an oil-operated valve with a movable outer sleeve controlled by a hydraulic servomotor. Water is discharged as a cone, usually into the tailwater. Figure 3-35 shows another oil-operated variant.

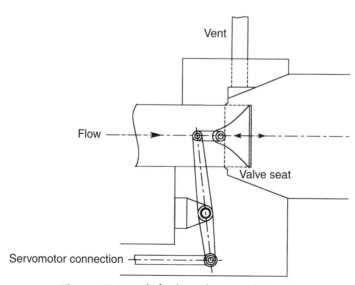

**Figure 3-34** Relief valve, oil operated, Type C.

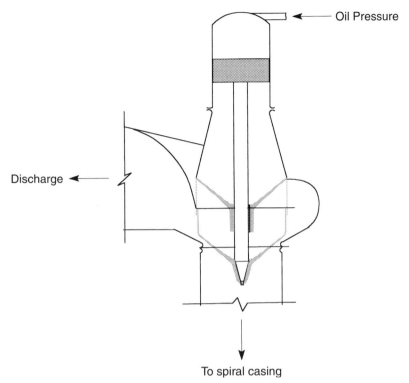

**Figure 3-35** Relief valve, oil operated, Type D.

# 3.8 Solved Problems

## Impulse Turbines

**3.8.1** Buckets of a Pelton wheel with a single nozzle are square to the nozzle jet when they are 10° from the vertical position as shown in Figure 3-4. The bucket speed is 0.47 × jet speed. Water leaves a bucket at 0.85 × incoming relative velocity. The supply head is 300 m above the nozzle, and the head lost in friction is 5% of the supply head. A single pipe with a Darcy-Weisbach friction factor of 0.03 supplies the nozzle. The nozzle has a $C_v = 0.97$ and a diameter = 3.2 cm. Find the power developed from the wheel and the necessary diameter of the pipe.

### Solution

From the velocity vector diagram in Figure 3-4 and noting that the bucket angle is 10° and not 15°:

$$w_2 = 0.85\, w_1 \tag{3.33}$$

and

$$U_1/v_1 = 0.47 \tag{3.34}$$

In this case $U_1 = $ component of the bucket speed U, parallel to the jet.

The overall efficiency:

$$\eta_0 = [(U_1/g)(w_1 + w_2 \cos 15)]/(v_1^2/2g) = (2U_1/v_1^2)(v_1 - U_1)(1 + 0.85 \cos 15°)$$
$$U_1 = U \cos 10° \tag{3.35}$$

Substituting values: $\eta_0 = 0.905$. Applying the Bernoulli equation from the beginning of the pipe to the nozzle:

$$H = h_f(\text{pipe}) + \text{nozzle loss} + v_1^2/2g \tag{3.36}$$

Substituting values:

$$300 = (0.05)(300) + (1/0.97^2 - 1)v_1^2/2g + v_1^2/2g$$
$$\therefore \quad v_1^2/2g = 268.2 \text{ m} \quad \text{and} \quad v_1 = 72.5 \text{ m/s}$$

The useful head from the wheel:

$$H_W = \eta_0 \times v_1^2/2g = (0.905)(268.2) = 242.7 \text{ m}$$
$$\text{Mass flow rate: } (1000)(\pi/4)(0.032^2)(72.5) = 58.3 \text{ kg/s} \tag{3.37}$$
$$\text{Power developed: } (58.3)(9.81)(242.7) = 138.8 \text{ kW}$$

**3.8.2** Typically, in a Pelton wheel the losses due to friction on the bucket and shock losses can be expressed as:

$$(k_1/2g)(v_1 - U)^2 \tag{3.38}$$

and the loss due to bearing friction as:

$$(k_2/2g)(U)^2 \tag{3.39}$$

where $k_1$ and $k_2$ are constants.

Show that the maximum efficiency occurs when:

$$(U/v_1) = (1 - \cos\theta + k_1)/[2(1 - \cos\theta) + k_1 + k_2] \tag{3.40}$$

Tests on a Pelton wheel having a bucket angle of 165° (refer to Figure 3-4) were made at an overall efficiency, $\eta_0 = 80\%$, with $(U/v_1) = 0.47$. What are the values of $k_1$ and $k_2$, and what are the losses as a fraction of the jet energy?

## Solution

The change in relative velocity is:

$$w_1 + w_2 \cos(180 - \theta) \tag{3.41}$$

With no losses: $w_1 = w_2 = v_1 - U$
Velocity change is:

$$(v_1 - U)(1 - \cos\theta) \tag{3.42}$$

The work done per unit specific weight is:

$$(U_1/g)(v_1 - U)(1 - \cos\theta) \tag{3.43}$$

With losses, the work done per unit specific weight becomes:

$$(U_1/g)(v_1 - U)(1 - \cos\theta) - (k_1/2g)(v_1 - U)^2 - (k_2/2g)(U)^2 \tag{3.44}$$

The jet kinetic energy/unit specific weight $= (v_1^2/2g)$
∴ the overall efficiency, $\eta_0 = 2(n - n^2)(1 - \cos\theta) - k_1(1 - n)^2 - k_2 n^2$
where:

$$n = (U_1/v_1) \tag{3.45}$$

$\eta_0$ will be a maximum when:

$$2(1 - 2n)(1 - \cos\theta) - 2k_1(1 - n) - 2k_2 n = 0 \tag{3.46}$$

that is, when,

$$n = (U/v_1) = (1 - \cos\theta + k_1)/[2(1 - \cos\theta) + k_1 + k_2] \tag{3.47}$$

Substituting values $n = 0.47$: $\theta = 15°$, we obtain: $k_2 = 0.248 + 1.13 k_1$. Substituting values in the equation for $\eta_0$, we obtain: $k_1 = 0.243$; $k_2 = 0.522$.

$$\therefore \text{Friction} + \text{shock losses} / \text{Jet energy} = (k_1/2g)(v_1 - U)^2/(v_1^2/2g) = k_1(1 - n)^2$$

$$= 0.068 \text{ or } 6.8\%$$

$$\text{Loss due to bearing friction} = (k_2/2g)(U)^2$$

$$\text{Bearing friction/Jet energy} = k_2 n^2 = 0.115 \text{ or } 11.5\%$$

**3.8.3** A Pelton wheel is designed to run at 500 rpm: the head across the machine varies from 500 to 550 m. When operating at 500 m, the overall efficiency is 80%. The wheel/jet diameter ratio at this level of operation is 12. $C_V$ for the jet is 0.98. What should be the jet diameters at heads of 500 and 550 m?

## Solution

Assuming that the design is such that the wheel is operating close to its maximum efficiency, then the specific speed may be calculated from:

$$206/N_S = D/d = 12$$

$$\therefore \quad N_S = 17.2 \tag{3.48}$$

From Figure 3-5 the speed factor is found to be $\Phi = 0.455$.
The bucket velocity is:

$$U = \Phi(2gH)^{0.5} = (0.455)[(2)(9.81)(500)]^{0.5} = 45.1 \, \text{m/s} \tag{3.49}$$

$$\omega = (u)/(D/2) = (500/60)(2\pi) = 52.36 \, \text{rad/s} \tag{3.50}$$

$$\therefore \quad D = (2)(45.1)/(52.36) = 1.72 \, \text{m and } d = 1.72/12 = 0.143 \, \text{m}$$

The jet velocity is given by:

$$V_1 = C_V(2gH)^{0.5} = (0.98)[(2)(9.81)(500)]^{0.5} = 97.1 \, \text{m/s} \tag{3.51}$$

The volumetric flow rate:

$$Q = aV_1 = (\pi/4)(0.143)^2(97.1) = 1.56 \, \text{m}^3/\text{s} \tag{3.52}$$

Power produced by the wheel:

$$P = \gamma Q H \eta_0 = (9.81)(1000)(1.56)(500)(0.80) = 6.12 \, \text{MW} \tag{3.53}$$

When the head changes to 550 m at the same wheel speed, the speed factor changes.

$$\text{Thus, } \Phi = u/(2gH)^{0.5} = 45.1/(2 \times 9.81 \times 550)^{0.5} = 0.434 \tag{3.54}$$

From Figure 3-4 the specific speed is approximately $= 22$.

The diameter of the jet required for the new head is:

$$d = N_S D/206 = (22)(1.72)/20 = 0.184 \text{ m} \tag{3.55}$$

The new jet velocity is:

$$V_1 = C_V(2gH)^{0.5} = (0.98)[(2)(9.81)(550)]^{0.5} = 101.8 \text{ m/s} \tag{3.56}$$

The volumetric flow rate:

$$Q = aV_1 = (\pi/4)(0.184)^2(101.8) = 2.71 \text{ m}^3/\text{s} \tag{3.57}$$

Finally, the power produced is:

$$P = \gamma Q H \eta_o = (9.81)(1000)(2.71)(550)(0.80) = 11.70 \text{ MW} \tag{3.58}$$

## Comment

The reason for the marked increase in power is that the jet flow rate was almost doubled because of the marked change of the jet diameter. There are, of course, limits to the cross-sectional area changes that can be made to a jet. In addition, as the specific speed increases, the efficiency decreases rapidly for such turbines. In this example, the efficiency was assumed to remain constant for the range encompassed by the $N_S$ changes, but in practice even these fairly small $N_S$ changes would cause a decrease of efficiency.

# Radial-flow Turbines

**3.8.4** An inward-flow reaction turbine (i.e., a Francis turbine) has a guide-vane angle of 10°, the inlet angle to the runner blades is 100°, and the outlet angle is 15°. Figure 3-36 shows the arrangement.

The guide vanes and the runner vanes reduce the flow area to inlet and outlet by 15%. Runner dimensions are: OD = 1 m; ID = 0.75 m; entrance width = 10 cm; exit width = 27 cm. The pressure head $(p_1/\gamma)$ at inlet = 4 m and at outlet $(p_2/\gamma)$ = 2 m. The elevation difference between inlet and outlet may be neglected $(z_1 = z_2)$. Assume that losses across the runner are 15% of the work done per kg of water flowing. What is the power developed at the runner?

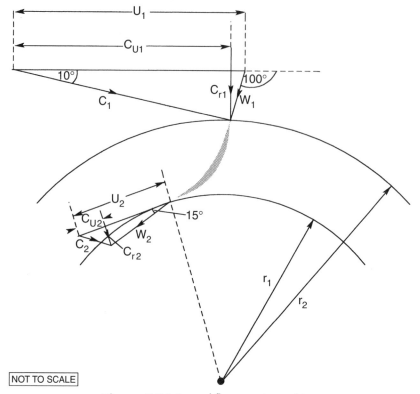

**Figure 3-36** Inward-flow reaction turbine.

## Solution

The available inlet runner area, $A_1 = 2\pi r_1 T_1(0.85) = 2\pi(0.5)(0.1)(0.85) = 0.27 \text{ m}^2$     (3.59)

Similarly, $A_2 = 2\pi r_2 T_2(0.85) = 2\pi(0.375)(0.27)(0.85) = 0.54 \text{ m}^2$     (3.60)

From continuity:

$$c_{r1}A_1 = c_{r2}A_2 \tag{3.61}$$

$$\therefore \quad c_{r2} = (0.27/0.54)c_{r1} = 0.5c_{r1} \tag{3.62}$$

From Figure 3-12:

$$(c_{r1})/(U_1 - c_{U1}) = \tan(80) \tag{3.63}$$

$$\text{Also:} \quad (c_{r1})/(c_{U1}) = \tan(10) \tag{3.64}$$

Combining Equations (3.63) and (3.64):

$$c_{U1} = 5.67\,c_{r1} \tag{3.65}$$

$$U_1 = 5.85\,c_{r1} \tag{3.66}$$

$$\text{Also,} \qquad U_2 = (r_2/r_1)U_1 = (0.375/0.5)(5.85\,c_{r1}) = 4.38\,c_{r1} \tag{3.67}$$

$$c_{r2} = 0.5\,c_{r1} = \tan 15\,(U_2 - c_{U2}) \tag{3.68}$$

$$\therefore \qquad c_{U2} = 2.51\,c_{r1} \tag{3.69}$$

The work done per unit-specific weight of water,

$$= (U_1 c_{U1} - U_2 c_{U2})/g = [(5.85\,c_{r1})(5.67\,c_{r1}) - (4.38\,c_{r1})(2.51\,c_{r1})]/g = 22.18\,c_{r1}^2/g \tag{3.70}$$

$$c_2^2 = c_{U2}^2 + c_{r2}^2 = (2.51\,c_{r1})^2 + (0.5\,c_{r1})^2 = 6.55\,c_{r1}^2 \tag{3.71}$$

The total head at runner exit is:

$$H_2 = p_2/\gamma + c_2^2/2g = 2 + 6.55\,c_{r1}^2/(2g) \tag{3.72}$$

The total head at runner inlet is:

$$H_1 = (\text{work done}) + (\text{runner loss}) + H_2 \tag{3.73}$$

$$\therefore \qquad H_1 = (44.36\,c_{r1}^2/2g) + 0.15(44.36\,c_{r1}^2/2g) + [(2 + 6.55\,c_{r1}^2)/(2g)] \tag{3.74}$$

Also,

$$H_1 = p_1/\gamma + c_1^2/2g = 4 + (c_{U1}^2 + c_{r1}^2)/2g = 4 + [(5.67\,c_{r1})^2 + c_{r1}^2)]/2g$$

$$= 4 + (33.15\,c_{r1}^2)/2g \tag{3.75}$$

$$\therefore \quad (44.36\,c_{r1}^2/2g) + 0.15(44.36\,c_{r1}^2/2g) + [(2 + 6.55\,c_{r1}^2/(2g)] = 4 + (33.15\,c_{r1}^2)/2g \tag{3.76}$$

Solving for $c_{r1}$ : $c_{r1} = 1.27\,\text{m/s}$

The specific weight per second flowing through the turbine is:

$$(dm/dt)\,g = \gamma\,A_1\,c_{r1} = (9.81)(1000)(0.27)(1.27) = 3364\,\text{N/s} \tag{3.77}$$

The power developed by the runner:

$$P = [(dm/dt)\,g](\text{work done}) = (3364)(44.36)(5.5)^2/(2g) = 230\,\text{kW} \tag{3.78}$$

**Comment**

The rotational speed of the runner $= 60/(\pi/U_1) = 142$ rpm.
The head across the turbine $= H_1 - H_2 = [4 + (33.15\,c_{r1}^2)/2g] - [2 + 6.55\,c_{r1}^2/(2g)] = 4.2\,\text{m}$.
Substituting in the equation for specific speed, $N_S = [(N)(P)^{1/2}/(H)^{5/4})] = 358$ (metric units). In the fl-lb-sec system this is 94. These numbers indicate that the turbine lies in the Francis turbine range. Furthermore, at the N calculated, the turbine should not be in a cavitating condition.

Making the flow exit radially, that is, causing the value of $c_{U2}$ to be $= 0$, could increase the power generated. In this case, the work done per unit-specific weight of water would increase from $22.18\ c_{r1}^2/g$ and become $= (U_1\ c_{U1})/g = [(5.85\ c_{r1})\ (5.67\ c_{r1})]/g = 33.17\ c_{r1}^2/g$.

**3.8.5** The following data were obtained on a turbine (the units are metric):

| Unit power, Pu | 10 | 10.5 | 10.7 | 10.7 | 10.5 | 10.0 |
|---|---|---|---|---|---|---|
| Unit speed, Nu | 50 | 55 | 60 | 65 | 70 | 75 |
| Mass flow rate, kg/s | 5395 | 5366 | 5273 | 5189 | 5147 | 5093 |

The values are plotted in Figure 3-37. The design head at maximum efficiency is 20 m. To what speed must the turbine be changed to operate at the same maximum efficiency if the head is changed to 25 m, and what power is developed at both heads?

**Solution**
Unit power is defined as:

$$Pu = P/H^{5/4} \tag{3.79}$$

$$\eta = 1000\,P/[(dm/dt)\,H] \tag{3.80}$$

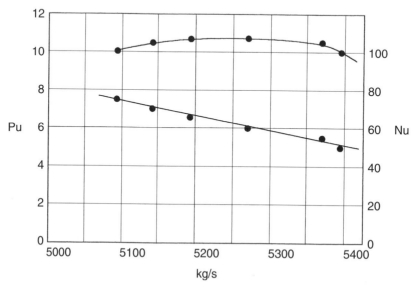

**Figure 3-37** Pu and Nu plotted as a function of mass flow rate.

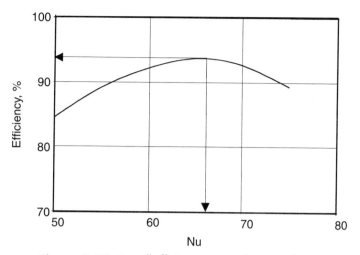

**Figure 3-38** Overall efficiency, % as a function of Nu.

where:

$P = kW, \quad h = m$

Combining Equations (3.79) and (3.80) yields:

$$\eta = 1000\, H^{1/2}\, Pu/(dm/dt) \tag{3.81}$$

Substituting the value of H and the value of Pu and dm/dt from the data:

| $\eta$, % | 84.5 | 89.2 | 92.5 | 94.0 | 93.0 | 89.5 |
|-----------|------|------|------|------|------|------|

These values are plotted in Figure 3-38.

It can be seen from Figure 3-38 that at the maximum overall efficiency of 94% the value of Nu = 66; the corresponding Pu = 10.7.

$$\therefore N_S = Nu(H)^{1/2} = (66)(10.7)^{1/2} = 216 \tag{3.82}$$

At the new head of 25 m; $N = (66)(25)^{1/2} = 330$ rpm
At a head of 20 m; $N = (66)(20)^{1/2} = 295$ rpm. The power developed at this speed is given by:

$$P = Pu\,(H)^{3/2} \tag{3.83}$$

Substituting values in Equation (3.83) for both heads: P(20 m) = 957 kW, P(25 m) = 1338 kW.

## Comment

As the head was increased, the rotational speed had to be increased. In practice, this may not be possible. If the speed had to be decreased, there would, of course, be a decrease in efficiency with a corresponding decrease in power.

### 3.8.6 Axial-Flow Turbines

An axial-flow turbine (see Figures 3-39 and 3-40) has the following geometric and flow data associated with it:

$$R_0 = 1 \text{ m} : R_H = 0.1 \text{ m} : R_T = 0.5 \text{ m}$$

$$b = 0.5 \text{ m} : N = 120 \text{ rpm}$$

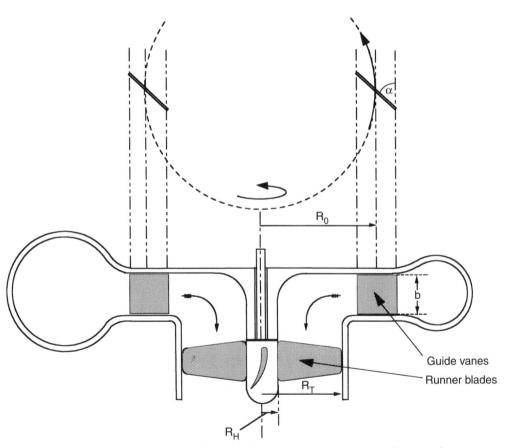

**Figure 3-39** Schematic diagram of the axial-flow turbine with plan view of two guide vanes.

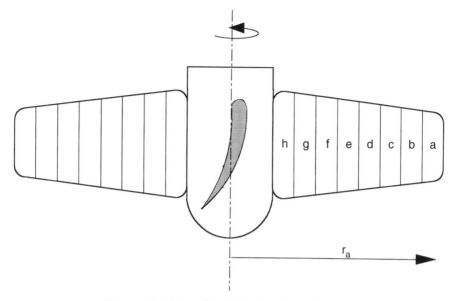

**Figure 3-40** Impeller divided into flow elements.

Effective head, H = 20 m

Guide-vane angle $\alpha = 15°$

Absolute velocity into guide vanes $V_0 = 10$ m/s

Assume that the fluid entering the draft tube has no angular momentum and that all losses may be neglected. Determine the ideal power developed.

**Solution**

The volumetric flow rate leaving the guide vanes and entering the runner is:

$$Q = 2\pi R_0\, bV_0 \sin \alpha = (2\pi)(1)(0.5)(10)(\sin 15°) = 8.13 \text{ m}^3/\text{s} \tag{3.84}$$

The axial velocity is:

$$V_{\text{axial}} = Q/\pi(R_T^2 - R_H^2) = 8.13/\pi(0.5^2 - 0.1^2) = 10.78 \text{ m/s} \tag{3.85}$$

Using the Euler turbine equation:

$$gH = (u_1 c_{u1} - u_2 c_{u2}) \tag{3.86}$$

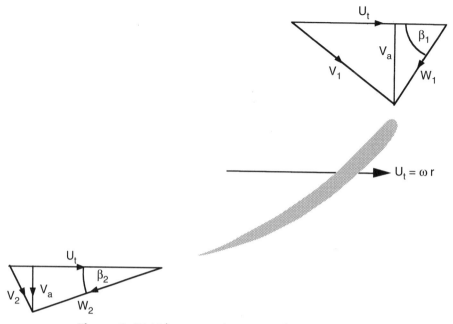

**Figure 3-41** Velocity triangles at any arbitrary position, r.

$c_{u2} = 0$ because the fluid has no angular momentum in the draft tube

$$\therefore \quad gH = u_1 c_{u1} = r \omega c_{u1}$$

$$c_{u1} = [(9.81)(20)]/[(2)(2\pi)r] = 7.8/r \tag{3.87}$$

A typical blade is for convenience divided into eight elements: a, b, c, d, e, f, g, h. The width of each element is $(0.5 - 0.1)/8 = 0.05$ m. Figure 3-41 shows velocity triangles at any arbitrary position, r.

The radius to the outer element a is thus: $r_a = 0.5 - 0.05 = 0.45$ m.

The fluid motion within the impeller is assumed to be a free vortex. This has been found to give good representation of the real velocity profile (see Chapter 2).

Thus:

$$(c_{u1})_{r(a)} = 7.8/r_a = 7.8/0.49 = 15.9 \text{ m/s}$$

The Euler turbine equation (Section 2.5) may now be used to calculate the torque on element (a).

$$M = (\gamma/g)Q(r_1 c_{u1} - r_2 c_{u2}) \tag{3.88}$$

In this case,

$$M = (\gamma/g)Q(r_1 c_{u1}) = (\rho)(Q)(r_1 c_{u1}) \tag{3.89}$$

For element (a),

$$(\Delta M)_a = (1000)(2\pi r_a)(0.05)(r_a)(7.8/r_a)(10.78) = 26{,}416\, r_a \qquad (3.90)$$

The power generated from this element is:

$$(\Delta P)_a = (\Delta M)_a \omega = (26{,}416)(0.45)(120/60)(2\pi) = 149 \text{ kW} \qquad (3.91)$$

The total power generated is:

$$\Sigma(\Delta P)_a = \Sigma(\Delta M)_a = 331.95(0.45 + 0.40 + 0.35 + 0.30 + 0.25 + 0.20 + 0.15 + 0.10)$$
$$= 730.3 \text{ kW}$$

**3.8.7** A large Kaplan-type turbine for a hydroelectric power station has the following relevant data:

Rate of rotation: 72 rpm
Runner tip diameter: 8 m
Hub/tip ratio: 0.4

When the prototype was installed and tested, it was found that for a head across the machine of 9.7 m and a volumetric flow rate of 300 m$^3$/s the measured power output was 24.7 MW. The mechanical efficiency, $\eta_M$, of the set is 97%, and the efficiency of the alternator is 96%. Determine the hydraulic efficiency and, assuming a free vortex flow model, determine the velocity triangles at the tip and hub of the runner.

**Solution**

The hydraulic power is: $[(g)(1000)(9.7)(300)]/(10^6) = 28.5$ MW
The overall efficiency $\eta_0$ is thus: $\eta_0 = 24.7/28.5 = 0.867$
The hydraulic efficiency $\eta_H$ is: $\eta_H = 0.867/[(0.96)(0.97)] = 0.93$
From the Euler turbine equation:

$$\eta_0(gH) = u_1 c_{u1(tip)} = (0.867)(9.806)(9.7) = 82.5 \text{ J/kg} \qquad (3.92)$$

Note that $c_{u2\,(tip)}$ is assumed to be zero.
The velocity of the blade tip:

$$u_{TIP} = \pi D_{TIP}/(N\,60) = (3.14159)(8)/[(72)(60)] = 30.16 \text{ m/s} \qquad (3.93)$$

Similarly, the velocity of the hub:

$$u_{HUB} = \pi D_{HUB}/(N\,60) = (0.4)(u_{TIP}) = 12.06 \text{ m/s} \qquad (3.94)$$

$$c_{u1(tip)} = 82.5/u_{TIP} = 82.5/30.16 = 2.74 \text{ m/s}$$

Similarly,

$$c_{u1(hub)} = 82.5/12.06 = 6.84 \text{ m/s}$$

The axial-flow velocity, $c_{axial}$, is given by:

$$c_{axial} = Q/[(\pi/4)(D^2 - d^2)] = (300)/[(\pi/4)(8^2 - 3.2^2)] = 7.11 \text{ m/s} \qquad (3.95)$$

The velocity triangles for the tip and the hub may now be drawn. These are shown in Figure 3-42.

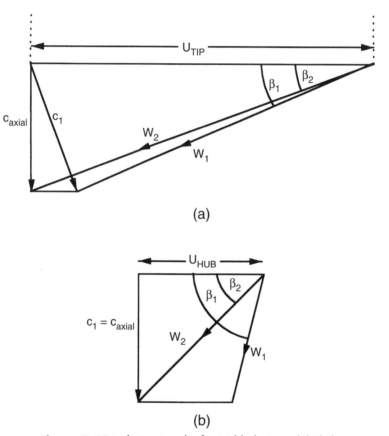

(a)

(b)

**Figure 3-42** Velocity triangles for (a) blade tip and (b) hub.

## Tip Angles:

$$\tan \beta_2 = c_{axial}/u_{TIP} = 7.11/30.16 = 0.2357: \quad \beta_2 = 13.3 \text{ degrees}$$

$$\tan \beta_1 = c_{axial}/(u_{TIP} - c_{u1(tip)}) = 7.11/(30.16 - 2.74) = 0.2593: \quad \beta_1 = 14.5 \text{ degrees}$$

## Hub Angles:

$$\tan \beta_2 = c_{axial}/u_{HUB} = 7.11/12.06 = 0.5896: \quad \beta_2 = 30.5 \text{ degrees}$$

$$\tan \beta_1 = c_{axial}/(u_{HUB} - c_{u1(hub)}) = 7.11/(12.06 - 6.84) = 1.3621: \quad \beta_1 = 53.7 \text{ degrees}$$

**3.8.8** A large Kaplan-type turbine with an output of 2600 kW operates at a net head that varies between 15 and 20 m. The optimal efficiency is at 20 m. Between 15 m and 20 m, the volumetric flow rate can be held constant, equivalent to an output of 2600 kW at 20 m.

## Solution

From Figure 3-22, C = 0.027.

∴ Runner diameter,

$$D_1 = 0.027(2600)^{0.5} = 1.38 \text{ m} \tag{3.96}$$

From Figure 3-21, specific speed,

$$N_S = N(P)^{0.5}/H^{5/4} = 450 \tag{3.97}$$

Substituting values in Equation (3.97):

$$N = (450)(20)^{5/4}/(2600)^{0.5} = 373 \text{ rpm} \tag{3.98}$$

The nearest synchronous speed is 360 rpm. The next step is to calculate a cavitation factor for a reasonably low suction head. At a net head of 20 m, the value of $\sigma$ is approximately 0.50 from Figure 10-3. The suction head is calculated from:

$$H_S = H_{ATM} - (\sigma H_{NET} + H_{VAP} + H_1) \tag{3.99}$$

where:

$H_{ATM}$ = atmospheric pressure
$H_{NET}$ = net head across turbine
$H_{VAP}$ = vapor pressure of water at prevailing temperature (Appendix A11)
$H_1$ = height of runner blade above centerline = $(0.15)(D_1)$

Substituting values for $H_{NET} = 20$ m:

$$H_S = 10.33 - [(0.5)(20) + (0.03 \times 10.2) + (0.15)(1.38)] = -2.94 \text{ m} \qquad (3.100)$$

Substituting values for $H_{NET} = 15$ m:

$$H_S = 10.33 - [(0.7)(15) + (0.03 \times 10.2) + (0.15)(1.38)] = -0.55 \text{ m} \qquad (3.101)$$

## Comment

These suction heads are acceptable. If the head fell below 15 m, the output and Q would have to be adjusted to make sure cavitation did not occur.

# 3.9 References

Daily, J.W., *Engineering Hydraulics*. Ed. H. Rouse, John Wiley & Sons, New York (1950), 943.

Daugherty, R.L., and Franzini, J.B., *Fluid Mechanics with Engineering Applications*. (6th Ed.). McGraw-Hill, New York (1965).

Gray, J.M., *Hydro-Electric Engineering Practice*. Ed. J.G. Brown, Blackie & Son Ltd., London (1958).

Mockmore, C.A., and Merryfield, F., The Banki Water Turbine. *Bulletin no. 25*, Engineering Experimental Station, Oregon State College, Corvallis (1949).

Quick, R.S., "Problems encountered in the design and operation of impulse turbines." *Trans. ASME*. 62 (1940).

*C H A P T E R    4*

# PUMPS

## 4.1 Introduction

Pumps and pumping systems have been known since the earliest times starting with the ancient Egyptians, Greeks, and Romans. They existed in one form or another for pumping water for irrigation and water supply to towns and cities. For many centuries to the end of the Dark Ages, development of the ancient inventions stagnated. However, the late eighteenth and early nineteenth centuries saw an increase in pump capabilities with the development of the steam engine.

The technological development of pumps accelerated rapidly as the power of the prime mover was increased; this was concomitant with the invention and development of the electric motor. The combination of these two machines not only revolutionized technology in the world, but produced such widespread types of pump that they have been used for almost an endless variety of services. It is fair to say that pumps are the second most common machines in the world after the electric motor.

As an illustration, a very broad classification of pumps is shown in Figure 4-1. In this figure two main streams are used in the classification: dynamic and displacement.

### 4.1.1 Theoretical Characteristics of Centrifugal Pumps

As a basis for understanding and differentiating theoretical behavior from the actual behavior of pumps it is necessary to distinguish the losses that are present. In Chapter 2 theoretical equations were presented for the theoretical head of pumps and turbines, the Euler equations. For a centrifugal pump, the relevant equation for the head generated is:

$$H_{TH} = (u_2 c_{U2} - u_1 c_{U1})/g \qquad (4.1)$$

where:

$_1$ and $_2$ refer to inlet and outlet
$u$ = tangential blade velocity
$c_U$ = tangential component of the absolute velocity c

**103**

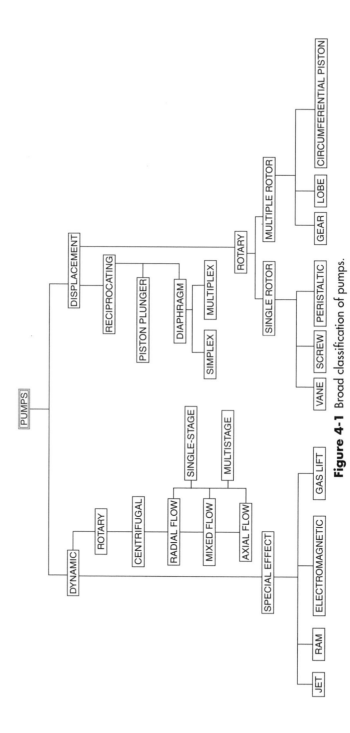

**Figure 4-1** Broad classification of pumps.

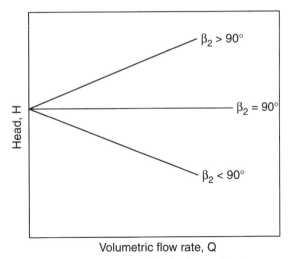

**Figure 4-2** H-Q relations for ideal centrifugal pumps.

The head predicted by Equation (4.1) is for ideal, that is, frictionless flow and an infinite number of blades, that is, no fluid slip. The head is also a function of the outlet angle of the blade $\beta_2$. Thus, for an ideal pump the head versus flow rate (H-Q) characteristics are a series of straight lines. When the fluid enters the pump radially, the $c_{U1}$ component is zero and Equation (4.1) becomes:

$$H_{TH} = u_2 c_{U2}/g \qquad (4.2)$$

The resulting H-Q relation is a horizontal straight line that is independent of Q. Figure 4-2 shows the relation between H and Q for different outlet angles $\beta_2$. It may be noted that when $\beta_2 > 90°$ the blades are forward-leaning, when $\beta_2 = 90°$ the blades are radial, and when $\beta_2 < 90°$ the blades are backward-leaning.

Losses that are inherent in any turbomachine cause the curves in Figure 4-2 to change in the manner illustrated in Figure 4-3.

## 4.2 Classification of Rotary Pumps

In this text, rotary pumps are classified into two broad categories:

1. Dynamic rotary—rotodynamic
2. Displacement rotary

The dynamic rotary pump includes all the variants of centrifugal and axial-flow types, and the displacement rotary pump includes vane, peristaltic, and other double rotor machines. Emphasis

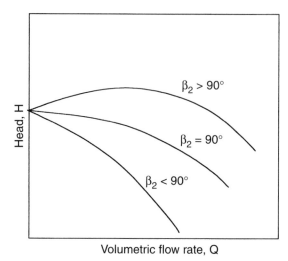

**Figure 4-3** H-Q relations for real centrifugal pumps.

will be placed here on the dynamic rotary, primarily because these machines are in a far greater preponderance than the displacement type.

Dynamic rotary or rotodynamic pumps are usually classified according to impeller shape. Shape and size markedly affect the value of specific speed, $N_S$. At the two ends of the spectrum are purely radial-flow pumps and purely axial-flow pumps. Shapes that are in between radial and axial flows are termed Francis or *mixed*. It should be noted that the term *mixed-flow pump* does not refer to the mixing of different streams but to the fact that the meridional velocity through the impeller has both radial and axial velocity components.

The simplest radial-flow pump we can think of has an impeller made up of a number of two-dimensional flat plates arranged symmetrically about a central point. The impeller may have a curved plate on the outside or may be open; the curved-plate impeller is referred to as *shrouded*, and the open as *unshrouded*. Flow, which must enter axially, is turned through 90° and exits radially from the impeller around the outer periphery; for an illustration, see Figure 4-4(a) and Figure 4-4(b). In Figure 4-4(a) the blades are flat; Figure 4-4(b) represents a slight variation of Figure 4-4(a). In this case, the blades have been curved backwards relative to the direction of rotation, but the curvature is only radial; there is no axial variation of curvature. Direction of curvature can have a profound effect on pump performance; for further details, see later in this chapter.

The properties of different impeller shapes, similar to turbines, are usually classified according to specific speed. They may be summarized as:

1. Low specific speed, $N_S < 1000$—radial-flow centrifugal pumps
2. Medium specific speed, $N_S = 1000–4000$—Francis-type impellers

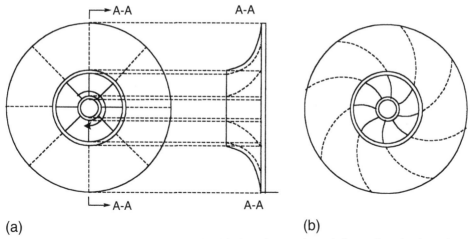

(a)                                              (b)

**Figure 4-4** (a) Simple radial impeller—flat blades; (b) simple radial impeller—backward-curved blades.

3. Medium to high specific speed, $N_S$ = 4000–10,000 mixed-flow types
4. High specific speed, $N_S$ > 10,000—axial-flow pumps

where specific speed was defined in Chapter 2 as $N_S = (NQ^{1/2}/H^{3/4})$.

$N_S$ in the above classification has units of N—rpm, Q—US gpm, and H—ft. These units are in common use in North America. For units in the SI system, that is, N—rpm, Q—m³/s, H—m, the above values must be multiplied by 0.01936. Figure 4-5 shows a general classification of single-entry centrifugal pumps and the effects of different impeller shapes on characteristic curves in terms of specific speed. Figure 4-6 shows a progression of pumps with diagonal impellers. In this diagram, the direction of the flow is becoming increasingly axial. At the extreme right of the diagram, Figure 4.6(d), the pump is referred to as a semi-axial-type pump. Figure 4.6(b) is the unshrouded version of Figure 4.6(a); similarly, Figure 4.6(d) is the unshrouded version of Figure 4.6(c).

A comparison of radial single-entry, radial double-entry, and mixed double-entry impellers is shown in Figure 4-7.

Figures 4-8 and 4-9 are schematics, typically showing the cross sections through radial pumps. These have backward-leaning impellers. Figure 4-8 shows a volute pump, and Figure 4-9 shows a diffuser pump, the diffuser blades surround the impeller. The inset diagram of Figure 4-8 shows details of the cutwater; the position of this has an important bearing on the performance of the pump and a number of other phenomena, which not only affect the characteristics of the pump, but may also lead to undesirable side effects such as resonance. The blades surrounding the impeller in Figure 4-9 serve to cause pressure recovery at outlet. This is accomplished by transforming the kinetic energy of the fluid into pressure energy.

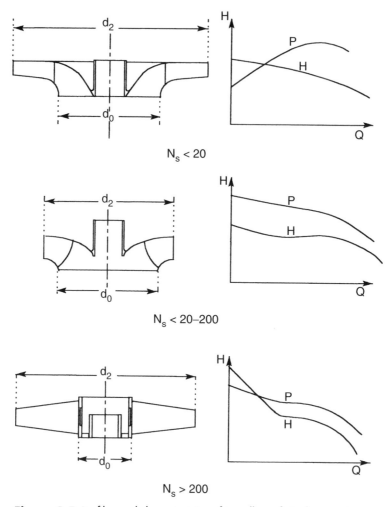

$N_s < 20$

$N_s < 20-200$

$N_s > 200$

**Figure 4-5** Profiles and characteristics of impellers of single-entry pumps.

# 4.3 Radial-flow Pumps

### 4.3.1 Geometry

Figure 4-10 illustrates the geometry of a typical cross section of a backward-leaning impeller; this represents a two-dimensional flow. In reality, the third component—normal to the plane of the figure—cannot be ignored. However, using this as a model produces a great deal of useful information. The velocity triangles at inlet and outlet are also shown. Absolute velocities are designated by c, tip velocities by u, and relative velocities along the blade by w. The velocity triangle construction at different radii may be seen in the figure. The u-components are always

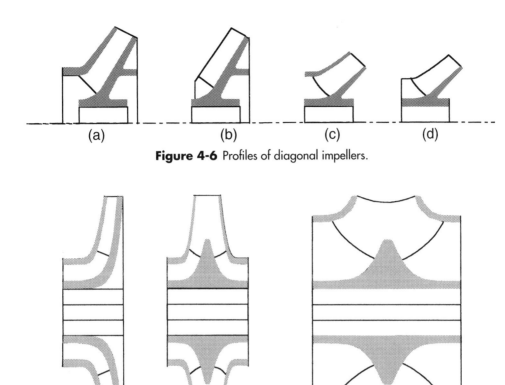

**Figure 4-6** Profiles of diagonal impellers.

**Figure 4-7** Comparison of (a) radial single-entry, (b) radial double-entry, and (c) mixed double-entry impellers.

tangential to circles centered at the shaft, the w-components are tangential to streamlines at any point in the flow, and the c-components are the vector addition of u and w. The components of the velocity triangles are broken down into more detail in Figure 4-11. Each streamline in each chamber, bounded by neighboring blades, is part of a conformal grid of which the blades themselves are part. If the flow were frictionless without any other losses, the fluid at outlet would be perfectly guided and the relative outlet velocity, $w_2$, would be tangential to the blade making an angle $\beta_2$ with the tip outlet velocity vector, $u_2$.

A common shape used for radial impellers is the logarithmic spiral, the graphical construction of which is shown in Figure 4-12. In this geometry, the blade angle is constant in the manner shown. Archimedean spirals are also used, and the graphical construction of one of these spirals is shown in Figure 4-13. In this spiral, equal angles are traced out at equal increments of radius.

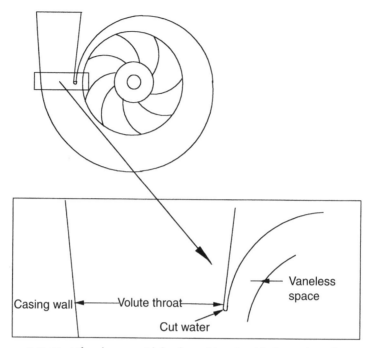

**Figure 4-8** Centrifugal pump with backward-leaning blades and a volute casing.

**Figure 4-9** Centrifugal pump with backward-leaning blades and a diffuser casing.

## 4.3.2 Power

The Euler equation for pumps (see Chapter 2, Section 2.5) is:

$$M = (\gamma/g)Q(r_2 c_2 \cos \alpha_2 - r_1 c_1 \cos \alpha_1) \tag{4.3}$$

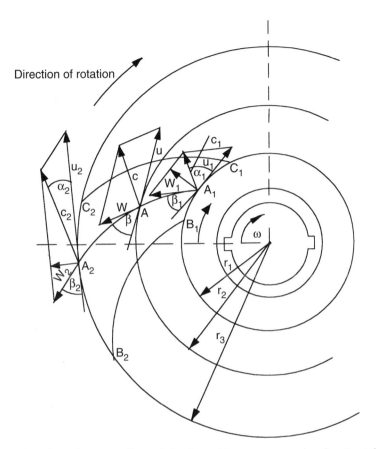

**Figure 4-10** Flow through an impeller with backward-leaning vanes also showing inlet and outlet velocity triangles $A_1$–$A_2$: median streamline $B_1$–$B_2$; $C_1$–$C_2$: neighboring blades $\beta_1 =$ inlet angle: $\beta_2 =$ outlet angle: $\alpha_1 = \sphericalangle (c_1, u_1) : \alpha_2 = \sphericalangle (c_2, u_2)$.

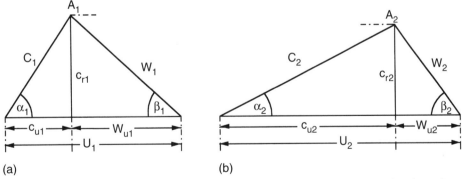

**Figure 4-11** Detailed velocity triangles of the impeller of Figure 4-10: (a) inlet; (b) outlet.

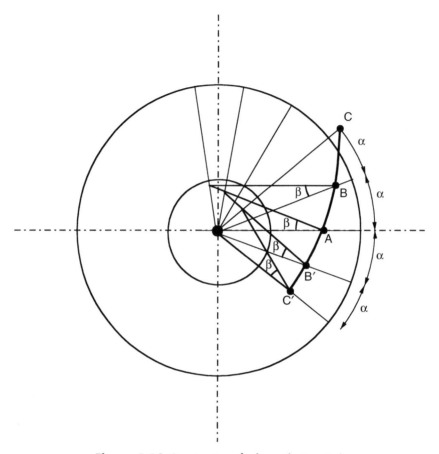

**Figure 4-12** Construction of a logarithmic spiral.

Equation (4.3) enables us to determine the hydraulic power needed by the pump. From the velocity triangles of Figure 4-11 it may be seen that:

$$c_{r1} = c_1 \sin \alpha_1; \; c_{r2} = c_2 \sin \alpha_2 \quad \text{and} \quad c_{u1} = c_1 \cos \alpha_1; \; c_{u2} = c_2 \cos \alpha_2 \tag{4.4}$$

The power transmitted to the liquid is:

$$P = M\omega = (\gamma/g)Q(r_2 c_{u2} - r_1 c_{u1})\omega$$

$$\omega = \text{angular rate of rotation} \tag{4.5}$$

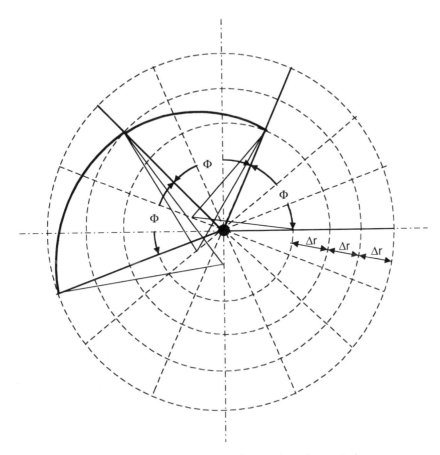

**Figure 4-13** Construction of an Archimedean spiral.

Referring to Figure 4-11, we see that:

$$P = (\gamma/g)Q(c_2 u_2 \cos \alpha_2 - c_1 u_1 \cos \alpha_1) \qquad (4.6)$$

Equation (4.6) may also be written as:

$$P = (\gamma Q)[(c_2^2 - c_1^2)/2g + (u_2^2 - u_1^2)/2g + (w_1^2 - w_2^2)/2g] \qquad (4.7)$$

From the Bernoulli equation we see that:

1. $(c_2^2 - c_1^2)/2g$: the increase in kinetic energy of the liquid
2. $(u_2^2 - u_1^2)/2g$: the energy expended in causing circumferential flow
3. $(w_1^2 - w_2^2)/2g$: change in relative energy form inlet to outlet—usually positive

### 4.3.3 Theoretical Head

The theoretical head $H_{th}(\infty)$ is the head which a pump with an infinite number of blades would generate in the absence of hydraulic losses and mechanical friction. Thus, for the flow of an ideal liquid through an ideal pump:

$$T_\omega = \gamma Q H_{th}(\infty)$$

$$T = \text{torque} \tag{4.8}$$

$$\omega = \text{angular rotational speed}$$

The theoretical head may be written:

$$H_{th}(\infty) = (1/g)(u_2 c_2 \cos \alpha_2 - u_1 c_1 \cos \alpha_1) \tag{4.9}$$

Note that the theoretical head is *independent of the liquid characteristics*. The head given by Equation (4.9) is the head given by a pump with an infinite number of blades. An infinite number of blades means, in effect, that there is no interblade circulation. The actual head developed will be dependent on the number of blades. Some authors prefer to use a definition of theoretical head or "virtual head" designated as $H_{virtual}(\infty)$, so that Equation (4.9) defining $H_{th}(\infty)$ requires modification for a finite number of blades. The phrase "finite number of blades" means that there is fluid circulation or a relative eddy in each chamber of the pump. This represents lost energy. $H_{th}(\infty)$ may be corrected to account for this by multiplying $H_{th}(\infty)$ by a factor called the slip factor $\mu$. Reference should be made to Section 4.2.8 for a discussion of slip factor and the equations proposed for this. Thus, for a finite number of blades:

$$H_{th}(z) = \mu H_{th}(\infty)$$

$$z = \text{blade number} \tag{4.10}$$

### 4.3.4 Energy Losses

Losses for any rotodynamic pump may be divided into four groups:

1. Head losses
2. Leakage losses
3. Disk friction loss
4. Mechanical losses

### 4.3.5 Head Losses

This group of losses includes frictional losses, contraction, expansion, and directional change losses. For the majority of machines, with the exception of small machines with passages that are small relative to the pump dimensions, these losses amount to less than 10%.

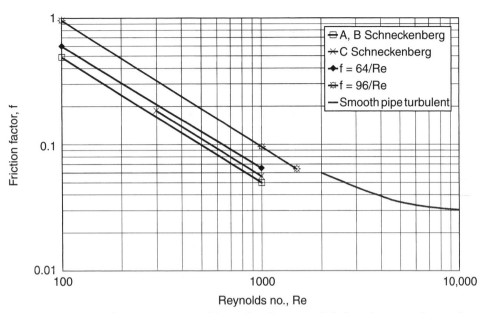

**Figure 4-14** Friction factor versus Reynolds number for a parallel plate slot. Smooth pipe flow is shown for comparison. (Modified from Stepanoff, 1957, courtesy John Wiley & Sons, Inc.)

The total head loss may be further broken down into:

1a. Entrance or profile losses

These losses are a function of the shape of the leading edge of the impeller blade. The blunter the blade, the greater the energy loss. The drag coefficient of any body that is traveling through a fluid or that has a fluid traveling past it is a function of its projected area in the flow direction and the shape of the leading edge. Any blade, if properly designed, should perform as an efficient wing section. This is as true for pump impeller blades as it is for turbine blades.

1b. Friction losses in the impeller passages

Modeling the flow as a liquid flowing through a parallel plate slot approximates the energy loss for a liquid flowing through such clearances. Stepanoff (1957) has presented experimental data on friction factors as a function of Reynolds number (see Figure 4-14). Notice that the friction factor is lower in laminar flow than either that predicted by infinite parallel plate flow or pipe flow. In turbulent flow, the values are about the same as in pipe flow.

## 4.3.6 Leakage Losses

Leakage may be divided into internal and external leakage. Internal leakage corresponds to that fraction of the fluid that recirculates inside the machine that is associated with the impeller.

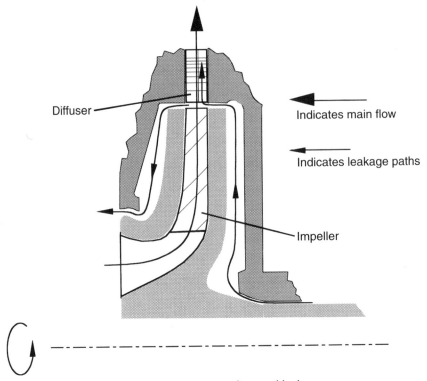

Diffuser

Impeller

Indicates main flow

Indicates leakage paths

**Figure 4-15** Illustration of internal leakage.

The flow for this leakage is through wearing rings, bushings, and balancing devices. External leakage is through stuffing boxes or outer seals. As a percentage of the total leakage, this is very small and may be neglected.

The impeller of a centrifugal pump rotates freely on its shaft and the clearances between the front and back of the impeller and the casing (see Figure 4-8) the cutwater must be such as to allow this to happen. Yet at the same time, because of these gaps the leakage must be minimized so that leakage is not excessive. In addition, there may be other problems with the cutwater, such as acoustic resonance. The values of these clearances are therefore a compromise between sealing and minimizing frictional losses and not creating resonance problems. Figure 4-15 illustrates the leakage paths at the front and rear of the impeller.

The leakage loss, for the purposes of obtaining a numerical estimate, may be regarded as:

$$Q_L = C_L \pi D_1 b_L (2gH_L)^{0.5} \tag{4.11}$$

where:

$C_L$ = a leakage coefficient varying from 0.3 to 0.7

$b_L$ = width of the leakage path

$H_L$ = head loss across the path

The leakage coefficient and the head loss are functions of the friction factor of the path, and the width of the leakage path depends on the tolerances of manufacture.

### 4.3.7 Disk Friction Loss

Frictional loss by a rotating disk has been the chief modeling tool for disk friction loss. Schultz-Grunow carried out a large number of experimental investigations on disk friction, which were correlated by Pfleiderer (1961). Pfleiderer defined the power absorbed by friction as:

$$P_{DF} = KN^3D^5 \qquad (4.12)$$

Equation (4.12) is an empirical equation for which K must be determined experimentally and is a function of the Reynolds number. Equation (4.12) appears to give conservative estimates of disk friction loss. A simpler equation, which is reliable from $N_S = 500 - 2000$ (rpm, US gpm, ft. units), has been suggested by Krutzsch (Karassik et al. 1976). It is:

$$(P_{DF}/P_W) = 7800/N_S^{5/3} \qquad (4.13)$$

where:

$P_W$ = waterpower

### 4.3.8 Mechanical Losses

Mechanical losses can only be calculated when the details of bearings and seals are known. The maximum value of loss as a percentage of hydraulic power value is only a few percent. Karassik et al. (1976) have presented an approximate relationship for the ratio of mechanical power loss to waterpower as a function of volumetric flow rate. The ratio decreases as the volumetric flow rate increases and as the specific speed of the pump increases. For example, at a flow rate of about 3 liters/s and a specific speed of 10, the ratio is 0.07. At a flow rate of about 600 liters/s and a specific speed of 100, the ratio is less than 0.006.

Stepanoff (1957) has also presented a useful series of power loss curves for various components of the flow for results obtained for double suction pumps. The components of power loss curve as a percentage of input power as a function of $N_S$ are shown in Figure 4-16. The total of these components is shown in Figure 4-17. Note that the units of $N_S$ are rpm, m, and $m^3/s$.

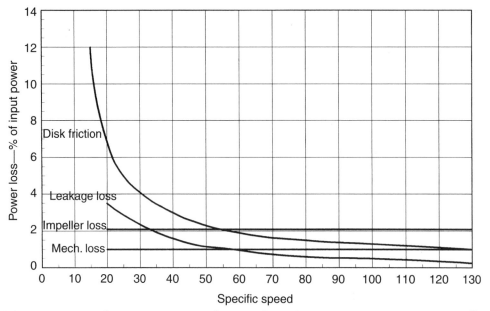

**Figure 4-16** Power loss components as a function of specific speed. $N_S$ units: rpm, m, and m$^3$/s. (Modified from Stepanoff, 1957, courtesy John Wiley & Sons, Inc.)

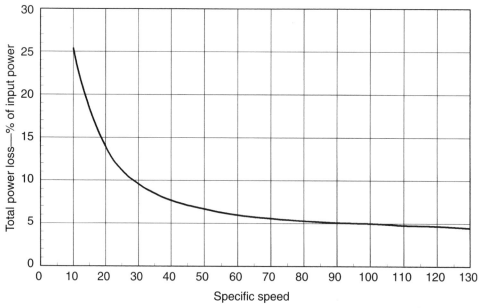

**Figure 4-17** Total power loss as a function of specific speed. $N_S$ units: rpm, m, and m$^3$/s. (Modified from Stepanoff, 1957, courtesy John Wiley & Sons, Inc.)

### 4.3.9 Specific Speed and Impeller Geometry

Referring to the velocity triangles of Figure 4-11, we may write, neglecting blade thickness:

$$Q = A_1 c_{r1} = \pi D_1 b_1 c_{r1} = \pi D_1 b_1 (u_1 - c_{u1}) \tan \beta_1 \qquad (4.14)$$

Assume that the fluid enters the pump radially that is, $c_{u1} = 0$

$$Q = \pi D_1 b_1 u_1 \tan \beta_1 = (\pi D_1 b_1)(\pi D_1 N/60) \tan \beta_1 \qquad (4.15)$$

Strictly speaking, there is never true radial entrance of the fluid. The presence of tangentially moving blades induces a prerotation in the entering fluid, so that $c_{u1}$ is not equal to zero. However, for our present purpose, it is small enough that it is a valid assumption. Hence:

$$H_{th}(\infty) = (1/g)(u_2 c_2 \cos \alpha_2 - u_1 c_1 \cos \alpha_1) \qquad (4.16)$$

$$\therefore \qquad H_{th} = H_{th}(z) = (1/g)(u_2 c_2 \cos \alpha_2) \qquad (4.17)$$

$$H = \eta_H H_{th} = \pi(\eta_H/g)u_2^2[1 - c_{r2}/(u_2 \tan \beta_2)] \qquad (4.18)$$

Substituting Equations (4.13) and (4.16) in $N_S = NQ^{1/2}/H^{3/4}$ and rearranging, we obtain:

$$N_S = (g/\mu\eta_H)^{3/4}(60/\pi^2)\{(D_1/D_2)[(b_1/D_2)\tan \beta_1]^{1/2}\}/[1-(D_1/D_2)^2(b_1/b_2)(\tan \beta_1/\tan \beta_2)]^{3/4}$$
$$(4.19)$$

To evaluate Equation (4.19) numerically, we assume that the velocity is constant through the impeller (no accelerations or decelerations), and we give average values of $\tan \beta_1$ and $\tan \beta_2$. In this case, the ratio ($\tan \beta_1/\tan \beta_2$) was assumed $= 0.7$. Equation (4.19) with these values becomes:

$$(N_S/\text{constant}) = [(D_1/D_2)(1/D_2^{1/2})]/[1 - 0.7(D_1/D_2)^3]^{3/4} \qquad (4.20)$$

Equation (4.19) was evaluated for a series of ($D_1/D_2$) ratios and values of $D_2$. These have been plotted in Figure 4-18. It can be seen that $N_S$ is a strong function of ($D_1/D_2$) and a weak function of $D_2$ alone. As ($D_1/D_2$) $\rightarrow$ 1.0 there is a marked increase in $N_S$ values. This is in agreement with the fact that propeller-type pumps have the highest value of $N_S$.

### 4.3.10 Modeling of Flow through an Impeller

The planar view of the flow through an impeller may be modeled by using a *Rankine* combined vortex together with a source flow. This is a solved problem at the end of Chapter 4. Looking laterally through the impeller, we see that the boundaries of the flow are the solid surfaces of

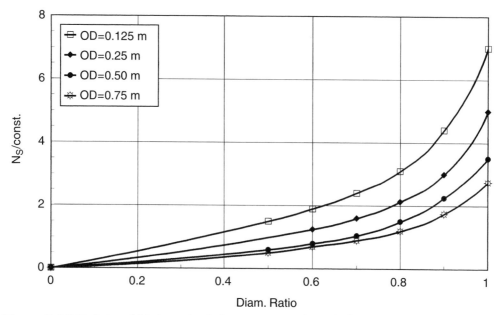

**Figure 4-18** Variation of $(N_S/\text{const.})$ with diameter ratio and outside diameter $(\tan\beta_1/\tan\beta_2) = 0.7$.

revolution as shown in Figure 4-19. This figure also shows one meridional streamline. The circles may be regarded as equal energy blocks. The flow may be divided into two components:

1. A meridional component $c_m$
2. A peripheral or circumferential component $c_u$

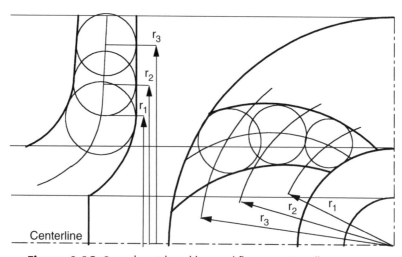

**Figure 4-19** Central meridional line and flow in an impeller passage.

It is assumed that due to axial symmetry the paths of the streamlines passing through the same circle are the same for all planes. The stream surfaces formed by contiguous streamlines are surfaces of revolution concentric with the impeller axis. However, the presence of a blade creates different flow conditions. The flow on the front (active) part of the blade is different from that on the rear (passive) part of the blade, which, in turn implies different pressures.

### 4.3.11 Axisymmetric Flow

In axisymmetric flow through an impeller, streamlines are axisymmetric surfaces of revolution lying around the center of rotation. The paths are determined by connecting the intersection of the meridional planes passing through the axis of the impeller. The velocity vectors tangential to the streamlines give the *meridional velocities*. In turbulent flow through the impeller, the streamlines coincide with the axisymmetric potential flow streamlines except close to the boundaries. The flow field is divided into a number of elements of equal volumetric flow. Figure 4-20 shows a passage with four elements. $\Phi$-lines must intersect with the streamlines, the $\Phi$-lines orthogonally. The result is a series of rectangles. The meridional velocity distribution is uniformly distributed across each element. The cross sections are surfaces of revolution, so that at each section the condition:

$$2\pi\,\text{rd} = \text{constant or rd} = \text{constant} \tag{4.21}$$

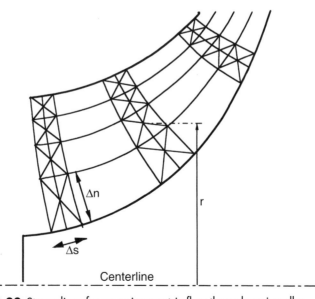

**Figure 4-20** Streamlines for an axisymmetric flow through an impeller.

Thus, the volumetric flow rate in an element of width d is:

$$\Delta Q = 2\,\pi\,\text{rd}\,C_M = \text{constant} \tag{4.22}$$

The velocity potential increases from element to element and is constant for individual paths. As the distance from the axis of rotation increases (i.e., as the passage becomes narrower), the rectangles formed by the intersection of the $\Phi$-lines and the $\Psi$-lines elongate in the direction of flow.

### 4.3.12 Net Positive Suction Head (NPSH)

An important factor in the installation of any centrifugal pump is its position relative to the system that it services. The line on the suction side of the pump is below atmospheric pressure, and if the pressure falls below the vapor pressure of the liquid being pumped, then vapor bubbles will form (see Chapter 9). This phenomenon can have devastating effects not only on the operation of the pump in terms of its efficiency but also on the pump itself in terms of metal removal because of the implosive energy of the bubbles. Furthermore, it can cause fatigue failure of the seals themselves. The NPSH of a pump may be defined as:

$$\text{NPSH} = (p_a/\rho g - p_v/pg) - Z_s) \tag{4.23}$$

where:

    $p_a$ = local atmospheric pressure
    $p_v$ = vapor pressure of liquid being pumped
    $\rho$   = density of liquid being pumped
    $Z_s$ = total suction head

The value of NPSH given by Equation (4.21) must always be positive.

### 4.3.13 Slip Factors

Even under frictionless flow conditions, the guidance of the fluid at the outlet does not follow the outlet angle of the vanes. The difference between the theoretical tangential velocity for an impeller with an infinite number of vanes, and the actual velocity for an impeller with a finite number of vanes, is called the slip $\Delta\theta$. $\Delta\theta$ is defined for *frictionless flows* as:

$$\Delta\theta = c_{u2}(\infty) - c_{u2}(z) \tag{4.24}$$

Slip factor for *frictionless flows* is defined as:

$$\mu = c_{u2}(z)/c_{u2}(\infty) \tag{4.25}$$

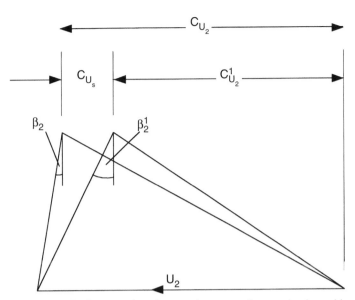

**Figure 4-21** Actual and theoretical velocity diagrams for a backward-leaning impeller. $\beta_2$ = theoretical outlet angle; $\beta_2^1$ = predicted actual flow outlet angle accounting for slip only.

In addition, when friction is taken into account, the slip is further increased. This means that the slip factor should be defined according to Equation (4.24) to account for all the losses contributing to an increased slip. Figure 4-21 illustrates the *real* (including friction and other losses) slip velocity for a pump:

$$c_{uS} = c_{u2} - c_{u2}^1 \qquad (4.26)$$

The slip factor for a real fluid is defined as:

$$\mu = c_{u2}^1 / c_{u2} \qquad (4.27)$$

The reason for the difference in frictionless flows is that a relative eddy exists in each chamber; that is, there is circulatory flow. The velocities generated here combined with the through flow velocities cause the fluid to appear to "slip." A model of what occurs is shown in Figure 4-22.

The difference between the predicted slip factor using Equation (4.23) and that predicted by Equation (4.25) is a function of Q, the volumetric flow rate. Numerous models have been hypothesized for the evaluation of the slip factor for frictionless flows. Stodola (1927) formulated two of the earliest expressions for slip and slip factor. For the slip velocity, he suggested the approximate relation:

$$c_{u2}(\infty) - c_{u2}(z) = u_2(\pi/z)\sin\beta_2 \qquad (4.28)$$

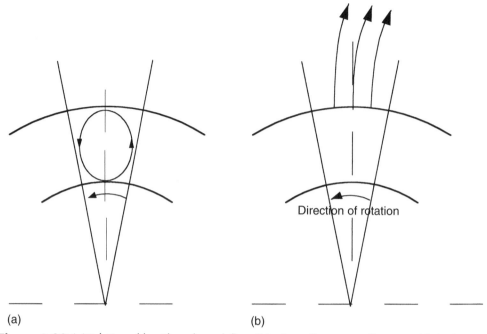

(a)                                          (b)

**Figure 4-22** (a) Relative eddy without through flow; (b) relative flow at impeller exit (added velocities).

An equation due to Stanitz (1952) is of simple form and is widely used for radial blades:

$$\mu = 1 - 0.63\,\pi/z \tag{4.29}$$

In Europe, a slip factor due to Pfleiderer (1961) is in widespread use; it is:

$$\mu = 1/\{1 + (a/z)(1 + \beta_2/60)[2/(1 - r_1^2/r_2^2)]\} \tag{4.30}$$

$a = 0.65$–$0.85$ for volute: $0.6$ for a vaned diffuser: $0.85$ to $1.0$ for a vaneless diffuser

Wiesner (1967) has reviewed and discussed slip factors at some length and has concluded that the slip factor due to Busemann (1928) that was originally developed for logarithmic spiral vanes has the greater validity over a wider range of outlet angles and blade numbers. The equation is:

$$\sigma_B = (A - B\Phi_2 \tan\beta_2)/(1 - \Phi_2 \tan\beta_2) \tag{4.31}$$

$$\Phi_2 = (c_{r2}/u_2)$$

A and B are functions of $(r_1/r_2)$, $\beta_2$, and z.

Further, if $(r_1/r_2) \geq \exp(2\pi \cos\beta^1/z)$, the function B may be taken to be 1.

Figure 4-23 shows a logarithmic spiral blade with the velocity components at a small element located along the blade, and Figure 4-24 shows the relationship between the function A, blade number, and $\beta_2$ for B $= 1$. Interpolated values may be made by the use of polynomial curve fits. The polynomials are:

$$A = 0.9: \beta_2 = -0.0011 \times (z^4) + 0.03262 \times (z^3) - 0.3413 \times (z^2) - 0.29292 \times (z) + 90.0$$
$$(4.32)$$

$$A = 0.8: \beta_2 = -0.02242 \times (z^4) + 0.40512 \times (z^3) - 2.6718 \times (z^2) + 1.83112 \times (z) + 90.0$$
$$(4.33)$$

$$A = 0.7: \beta_2 = -0.0417 \times (z^4) - 0.4167 \times (z^3) - 2.5417 \times (z^2) - 14.0833 \times (z) + 90.0$$
$$(4.34)$$

$$A = 0.6: \beta_2 = -7.5 \times (z^2) - 15.0 \times (z) + 90.0 \tag{4.35}$$

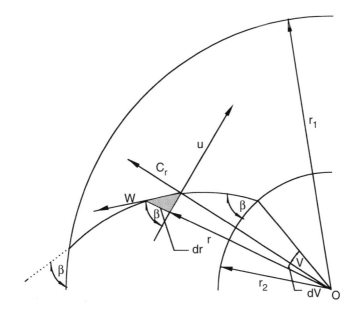

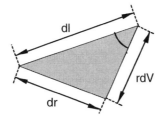

**Figure 4-23** Logarithmic spiral vane showing velocity components of an element.

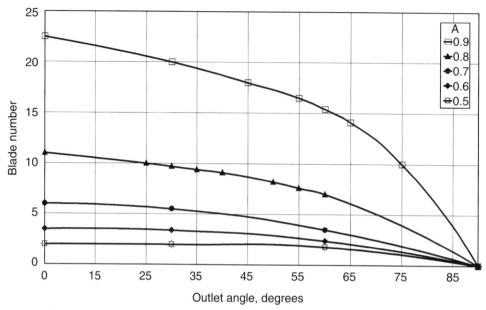

**Figure 4-24** Relationship between constant A, blade number, and $\beta_2$ for B = 1.

Finally, Figure 4-25 is a plot of head H, versus volumetric flow rate Q, showing the effects of all the losses discussed above on the theoretical head $H_{th}(\infty)$.

### 4.3.14 Effect of Blade Number, Outlet Blade Angle, and Circulation in Blade Passages

Varley (1961) studied the effects of blade number on pump performance. He used a machine with the following characteristics: H = 16 m, impeller diameter 244 mm, N = 1400 rpm. Higher outlet angles gave unstable characteristics. Figures 4-26, 4-27, and 4-28 summarize his investigations. Figure 4-26 shows the effects of $\beta_2$ varying from 15° to 88° on the relation between the head coefficient and the flow coefficient. Values of $\beta_2$ greater than about 27° gave unstable characteristics. This meant that the curves with $\beta_2 > 27°$ had maxima; this is consistent with pump surge occurring at values of flow coefficient greater than the value for the maximum. Similar behavior was encountered with blade number. Blade numbers above six gave instabilities (see Figure 4-27). Overall efficiency also peaked at about a blade number of six (see Figure 4-28). The overall effect is that a finite number of blades of finite thickness increases the angles $\alpha_1$ and $\beta_1$ at inlet and decreases the angles $\alpha_2$ and $\beta_2$ at outlet. The recommended designs had five to six blades with $\beta_2 = 27°$.

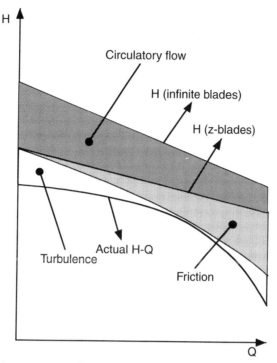

**Figure 4-25** Effect of losses on H(∞)-Q characteristics.

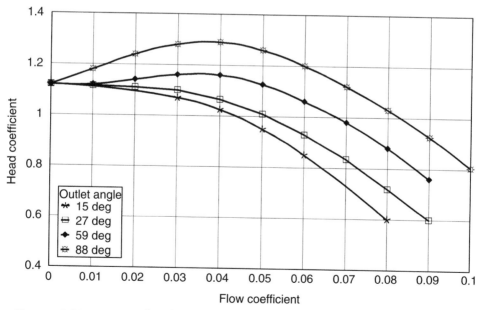

**Figure 4-26** Summary of Varley's (1961) results of the effect of $\beta_2$ on pump performance.

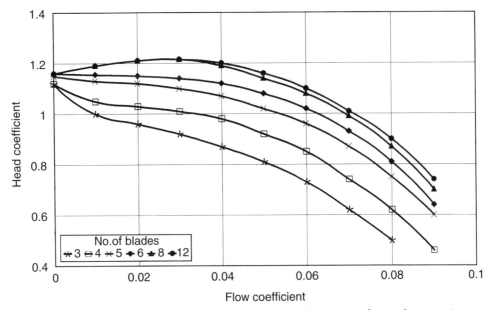

**Figure 4-27** Effect of number of blades on pump performance (after Varley, 1961).

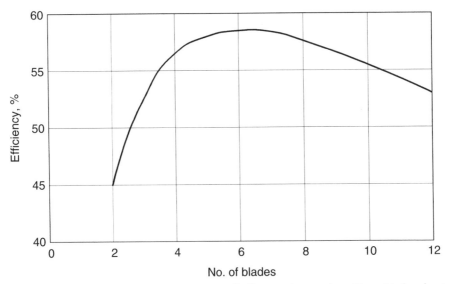

**Figure 4-28** Effect of blade number on overall efficiency (cross plotted from Varley data).

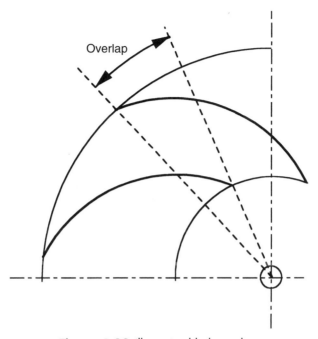

**Figure 4-29** Illustrating blade overlap.

### 4.3.15 Choice of Blade Number and Blade Overlap

Stepanoff (1957) quotes a blade number equation as:

$$z = \beta_2/3 \tag{4.36}$$

Another equation in common use is due to Pfleiderer (1961):

$$z = 6.5[(D_2 + D_1)/(D_2 - D_1)] \sin \beta_M \tag{4.37}$$

The definition of *blade overlap* is best illustrated in Figure 4-29. Experiments carried out with a series of impellers having five to nine blades have shown that overlap should be between 30° and 45°.

### 4.3.16 Energy Recovery

A common way of recovering pressure energy is to add a diffuser at the circumferential outlet of a radial-flow pump. This is illustrated in Figure 4-30.

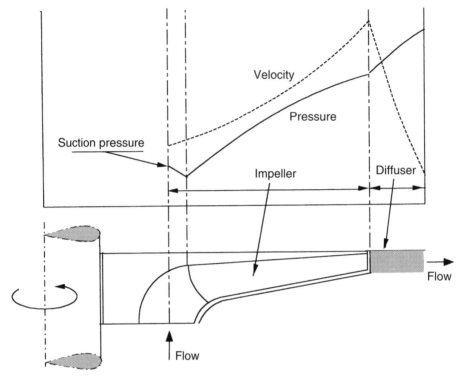

**Figure 4-30** Effect of a diffuser on the exit pressure and velocity of a radial pump.

### 4.3.17 Examples of Radial-flow Pumps

Figure 4-31 illustrates schematically a single-entry, overhung, radial-flow pump. Notice the hole through the back plate that enables partial pressurization of the rear surface, thus balancing part of the axial thrust.

Figure 4-32 is an example of a single-entry process pump with a radial impeller used in the petrochemical industry, with Figure 4-33 showing an example of a heavy-duty, double-entry pump.

### 4.3.18 Installation of a Typical Centrifugal Pump

Figure 4-34 illustrates a simple typical installation of one centrifugal, single-stage pump with suction lift, that is, negative head. The suction side could also be horizontal (i.e., at the same level of the centerline of the pump) or have a positive suction head (i.e., the suction level above the centerline of the pump).

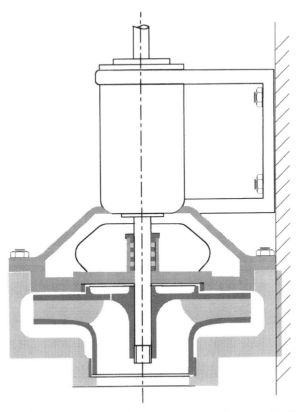

**Figure 4-31** Schematic example of a single-entry, overhung, radial-flow pump.

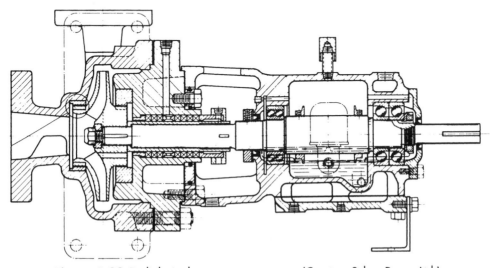

**Figure 4-32** Radial, single-entry process pump. (Courtesy Sulzer Pumps Ltd.)

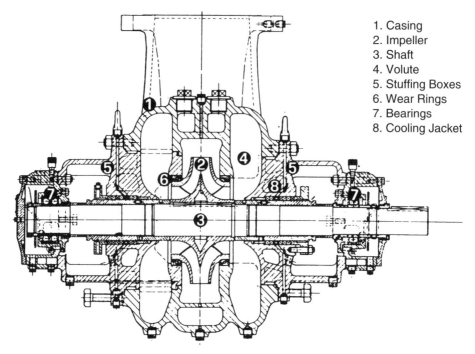

1. Casing
2. Impeller
3. Shaft
4. Volute
5. Stuffing Boxes
6. Wear Rings
7. Bearings
8. Cooling Jacket

**Figure 4-33** Heavy-duty, double-entry, radial-flow pump. (Courtesy Goulds Pumps Inc.)

## 4.3.19 Special-purpose Radial-flow Pumps

There are many variants on the radial centrifugal pump: these are designed specifically for difficult locations or handling specific materials. Pump operation may require horizontal or vertical positions and partial or complete submergence with automatic self-priming either internally or externally. Fluid systems may be corrosive liquids, waste liquids, and liquids containing fibrous and particulate solids. Process pumps might have to operate at high temperatures; also, process pumps and certain pumps might have to be entirely isolated from the fluid (e.g., "canned" pumps).

Figures 4-35(a) and (b) are examples of impellers designed for liquid-solid systems, that is, nonclogging impellers. They are designed in such a way that the fluid, usually containing particle or fibers that may stick together or intertwine, is kept in a state of highly turbulent agitation, while in contact with the impeller.

Figure 4-36 illustrates a commercial pump for paper stock pumping. This figure shows that the input configuration has been specifically designed for fibrous materials. The suction end incorporates a stirring mechanism followed by an Archimedean screw.

Figure 4-37 is a typical example of a "canned" pump. This type of pump is completely sealed, with the electric driver directly coupled to the pump. The pump can operate in a completely submerged liquid environment.

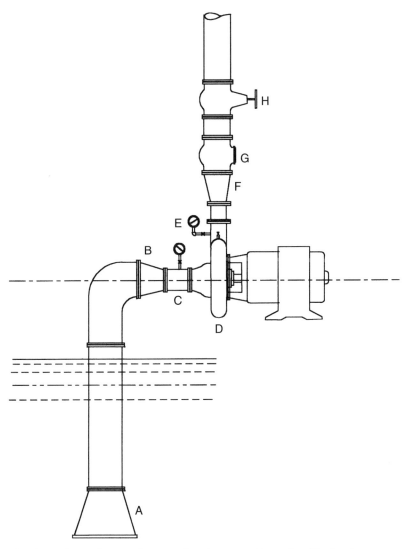

**Figure 4-34** Typical installation of a centrifugal pump with suction lift showing connection components: A—strainer; B—reducer; C—pressure gage inlet; D—pump; E—pressure gage inlet; F—expander; G—sight glass; H—gate valve.

## 4.4 Mixed-flow Pumps—Diagonal Impeller Pumps

Mixed-flow pumps are used when a higher specific speed than that normally associated with radial flow pumps is required. Figures 4-38 and 4-39 show two types of mixed-flow pumps.

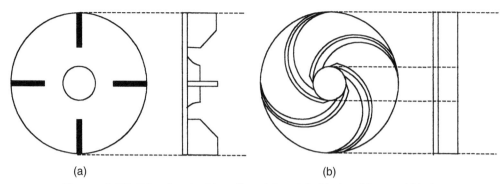

(a)                                                    (b)

**Figure 4-35** "Nonclogging" impellers: (a) straight blades; (b) curved blades.

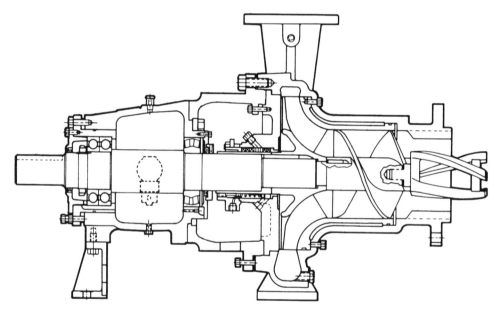

**Figure 4-36** An example of a pump for paper stock or waste pumping. (Courtesy Goulds Pumps Inc.)

## 4.5 Axial and Semiaxial Pumps

A typical *axial-flow* pump is shown in Figure 4-40, together with the velocity vector diagrams for impeller blade and diffuser blade in Figure 4-41.

It is usual to plot velocity diagrams on one plot for convenience, with the inlet and outlet velocity vectors having a common base. The common base is the blade velocity u, and it is easier to see the effects of angle and velocity vector changes in this way (see Figure 4-42). For example, if the tangential velocity component at outlet $c_{U2}$ was needed to be reduced to zero in order to

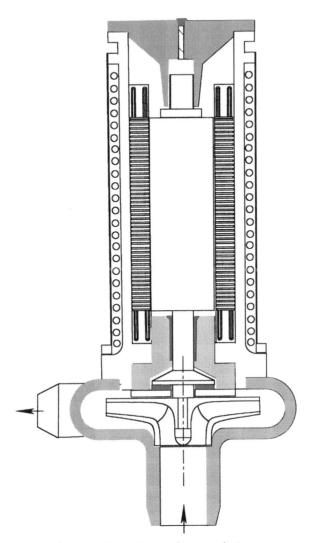

**Figure 4-37** A "canned" pump design.

increase the efficiency, then $c_2$ would become identical to the axial velocity, $c_a$, and angle $\beta_2$ would become less than angle $\beta_1$. There are, of course, other constraints on the amounts by which such changes can be made, such as blade speed.

Semiaxial pumps may have fixed or rotatable blades. They are diagonal-flow pumps with their blades usually projecting forward in the axial direction. Figure 4-43 shows a side and end elevation of a semiaxial pump, with Figure 4-44 showing a three-dimensional view of a typical impeller of this sort.

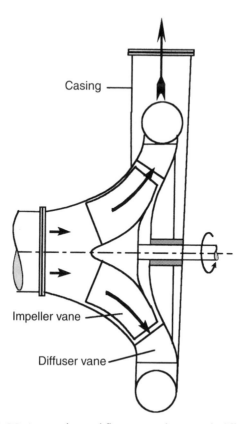

**Figure 4-38** A typical mixed-flow pump design with diffuser blades.

The *Deriaz* pump, an impeller of which is illustrated in Figures 4-45(a)–(c), is the pump counterpart of the *Deriaz* turbine. It is a semiaxial pump with rotatable blades. Figure 4-45(a) is a plan view of the impeller showing the blades closed; Figure 4-45(b) is the same view with the blades fully open, that is, allowing the flow to be maximum. Figure 4-45(c) is a side elevation of the pump showing the position of a blade. A distinct advantage of the *Deriaz* pump is that it allows good flow control.

### 4.5.1 Unbounded Axial Impellers or Propellers

The extension of the theory and application of all that has been discussed in Section 4.4 to an unbounded axial impeller or propeller should be self-evident. The equations to be derived in this section will be equally applicable to air flows, that is, aerodynamics. We first consider the unbounded flow past a propeller as shown in Figure 4-46.

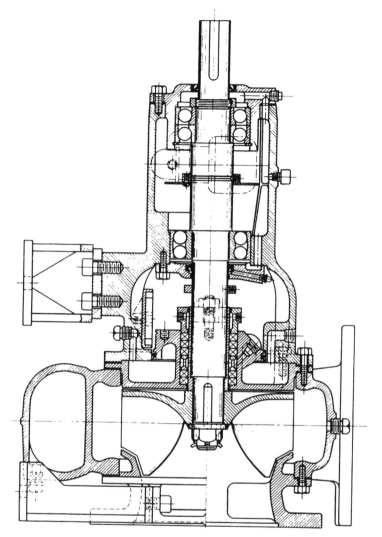

**Figure 4-39** A commercial mixed-flow pump designed for chemical service.

The control surface used in this analysis is the surface bounded by the top and bottom stream-lines and the upstream and downstream parallel streams of Figure 4-46. A momentum balance on left and right faces results in an axial force $F_A$, given by:

$$F_A = \rho Q(Va_4 - Va_1) = \rho A_{PROP}Va_0(Va_4 - Va_1) \qquad (4.38)$$

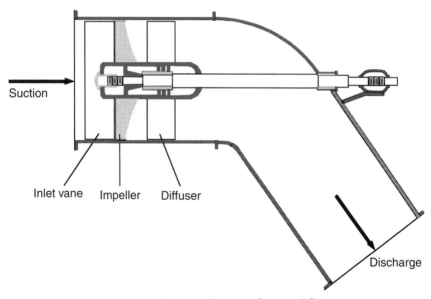

**Figure 4-40** A schematic diagram of an axial-flow pump.

where:

$\rho$ = fluid density

Q = volumetric flow rate

$A_{PROP}$ = flow area through the propeller

$$A_{PROP} = (\pi/4)(D_{TIP}^2 - D_{HUB}^2) \tag{4.39}$$

If the pressures immediately upstream and immediately downstream of the propeller are $p_2$ and $p_3$, then applying the Bernoulli equation between 1 and 2:

$$p_1 - p_2 = (\rho/2)(Va_2^2 - Va_1^2) \tag{4.40}$$

Similarly, applying the Bernoulli equation between 3 and 4:

$$p_3 - p_4 = (\rho/2)(Va_4^2 - Va_3^2) \tag{4.41}$$

The pressure distributions are shown in Figure 4-47.

We note that the far-upstream pressure is equal to the far-downstream pressure, that is, $p_1 = p_4$. Also, the axial velocity is very nearly the same before and after the propeller, that is, $Va_2 = Va_3$.

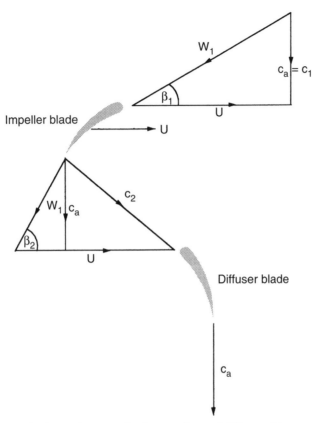

Impeller blade

Diffuser blade

**Figure 4-41** Typical velocity diagrams for the impeller and diffuser of the pump in Figure 4-40.

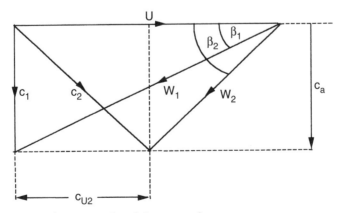

**Figure 4-42** Velocity triangles of diagrams of Figure 4-41 put on a common base.

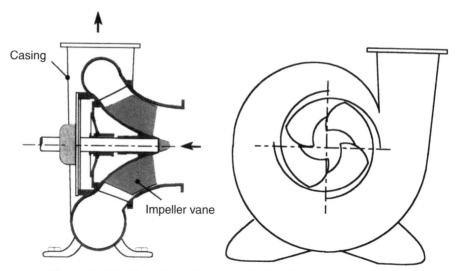

**Figure 4-43** Side and end elevations of a fixed-bladed semiaxial pump.

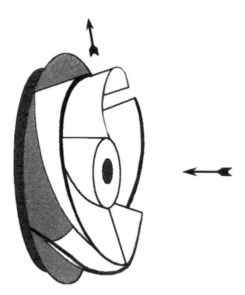

**Figure 4-44** Three-dimensional view of a fixed-bladed semiaxial pump impeller.

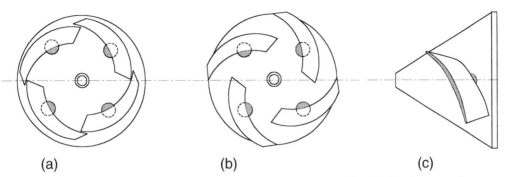

(a)                                    (b)                                    (c)

**Figure 4-45** Deriaz pump impeller: (a) plan view showing blades closed; (b) plan view showing blades fully open; (c) side elevation showing the position of a blade on the hub.

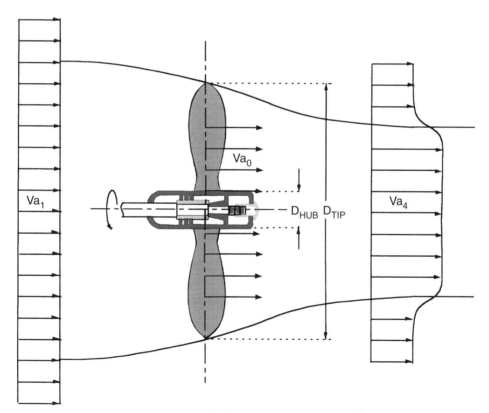

**Figure 4-46** Hydrodynamic flow past a propeller.

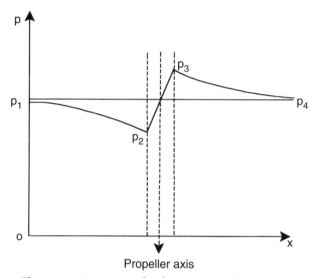

**Figure 4-47** Pressure distribution across the propeller.

With these conditions, Equations (4.40) and (4.41) when added together yield:

$$p_4 - p_1 = (\rho/2)(Va_4^2 - Va_1^2) \tag{4.42}$$

For frictionless flow, the power delivered to the propeller is identical to the change in kinetic energy of the fluid. Using Equation (4.41) in conjunction with Equation (4.42) gives:

$$Va_0 = (Va_4 - Va_1)/2 \tag{4.43}$$

The power supplied to the propeller is:

$$(dw/dt)_{INPUT} = \rho A_{PROP} Va_0 (Va_4^2 - Va_1^2)/2 \tag{4.44}$$

It is also useful to let $Va_1$ be defined as simply V and write:

$$Va_4 = Va_1 + \Delta V = V + \Delta V \tag{4.45}$$

Equation (4.44) with this slight, more convenient change of notation becomes:

$$(dw/dt)_{INPUT} = \rho A_{PROP} V \Delta V (1 + \Delta V/2V) \tag{4.46}$$

The useful power is:

$$(dw/dt)_{USEFUL} = \rho A_{PROP} V \Delta V \tag{4.47}$$

Thus, the propulsive efficiency is:

$$\eta_{PROP} = [(dW/dt)_{USEFUL}]/[(dW/dt)_{INPUT}] = 1/(1 + \Delta V/2V) \tag{4.48}$$

Equation (4.48) is of fundamental importance. It indicates that the efficiency of propulsion may be increased by increasing the velocity relative to the fluid. The difficulty here with marine applications is the possibility of the onset of cavitation. Cavitation in turbomachines is discussed in some detail in Chapter 9.

Performance characteristics of propellers are determined experimentally in water tunnels. Five dimensionless groups are of significance for such data:

The thrust coefficient:

$$C_{FA} = F_A/\rho N^2 D^4 \tag{4.49}$$

The torque coefficient:

$$C_T = T/\rho N^2 D^5 \tag{4.50}$$

The power coefficient:

$$C_P = P/\rho N^3 D^5 \tag{4.51}$$

Efficiency:

$$\eta = F_A V/\omega T \tag{4.52}$$

Speed of advance coefficient:

$$J = V/ND \tag{4.53}$$

## 4.6 Pump Characteristics of Centrifugal Pumps

### 4.6.1 Single Centrifugal Pumps—Radial- and Mixed-flow Impellers

Up to this point, we have seen how various losses have an effect on pump performance, that is, the H-Q curves for real pumps. Figures 4-48, 4-49, and 4-50 show characteristic curves for real pumps.

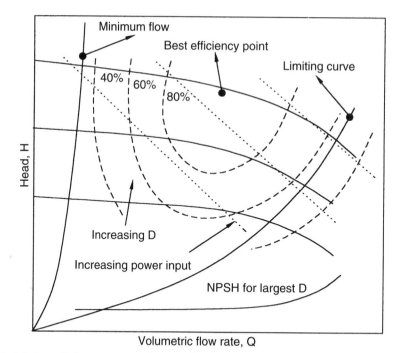

**Figure 4-48** Typical characteristics of a centrifugal pump with radial or backward-leaning blades.

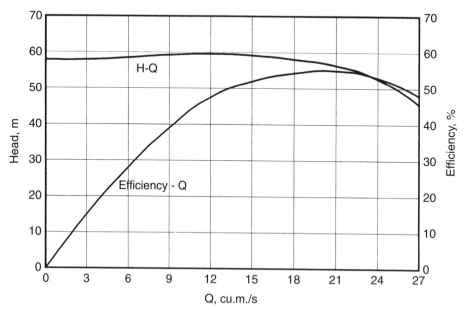

**Figure 4-49** Experimental data for a pump with radial blades. Inlet and outlet angles = 90°: OD = 13.3 cm.

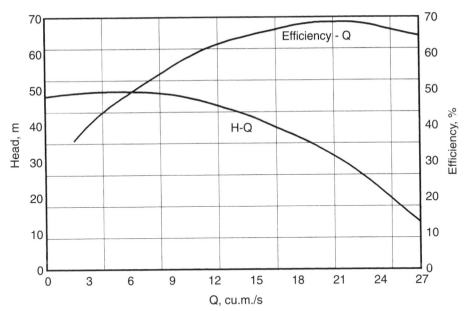

**Figure 4-50** Experimental data for a pump with backward-leaning blades. N = 1790 rpm. Inlet blade angle = 27°: outlet angle = 23°: ID = 22.2 cm: OD = 35.6 cm.

### 4.6.2 Effect of Fluid Properties

A change in fluid properties, such as density, viscosity, by the addition of solids in particulate form in the liquid (slurry), or by the introduction of air bubbles into the liquid can change the pump characteristics. Figure 4-49 illustrates the effects of real or apparent viscosity increase on efficiency, head, and power in a general way. As viscosity increases, the head required for a given flow rate increases, the efficiency decreases, and the power increases. The actual values of the curves will be functions of the type of pump and the rheology of the fluid. See Figure 4-51.

## 4.7 Series and Parallel Connections

Two or more pumps, either identical or nonidentical pumps, may be connected in series or in parallel to achieve a set of required flow conditions. The objective of a combination of pumps is to try and ensure that the operating point of the combination is close to the maximum efficiency of the combination. Cost is an overriding factor here. There is obviously no point in having a pump combination to replace a single larger pump if the capital and operating costs over the projected life of the system are greater, even though the overall efficiency of the combination is closer to the operating point. The effect of each pump on the combined characteristics is illustrated for a pair of identical pumps connected in series in Figure 4-52 and for nonidentical pumps connected in series in Figure 4-53. The combination in the latter case is not shown. However, in each case, the

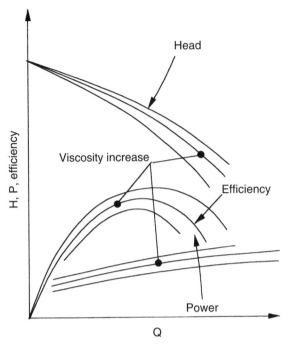

**Figure 4-51** A generalized diagram showing the effect on pump characteristics with change of fluid properties.

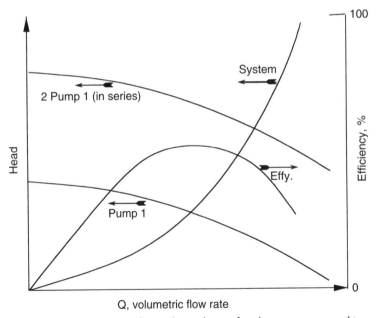

**Figure 4-52** Characteristics of two identical centrifugal pumps connected in series.

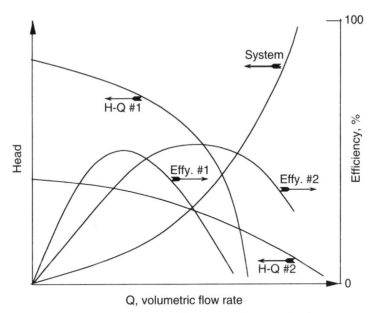

**Figure 4-53** Characteristics of two dissimilar centrifugal pumps.

heads of the pumps are added. When such pumps need to be connected in parallel, the volumetric flow rates are added. Examples of these connections are given in Section 4.9. The size and correct combination of pumps are dependent on the system curve and any demand changes that may occur. The aim in all of this is to have a pump combination that will meet the system requirements and exhibit the best efficiency over a wide range of conditions.

When pumping flow rate requirements vary, it may be advisable to connect a number of pumps in parallel. In this case if the flow rate demand declines, one or more pumps may be shut down, allowing the remaining working ones to operate at close to maximum efficiency. If one large pump is used, then if demand is lowered, the pump will have to be throttled with consequent reduced efficiency.

Figures 4-54, 4-55, and 4-56 show typical identical pump series connections. The aim of pumps connected either in series or parallel is to optimize the operation of the system, that is, to end up as close to the maximum efficiency as possible for changing conditions. The purpose of using a hydraulic coupling is to smooth out flow discontinuities between the pumps as a result of different characteristics. The system with no hydraulic coupling represented by Figure 4-54 is driven at constant speed, possibly resulting in very different efficiencies from each pump, causing a resultant overall efficiency of the system that could be far removed from the optimum. Of the three systems shown in these three figures, the one shown in Figure 4-56 is the best system to use for energy optimization.

A practical connection of the pumps shown in Figure 4-56 is shown in Figure 4-57.

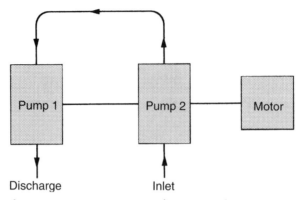

**Figure 4-54** Pumps connected in series—direct coupling.

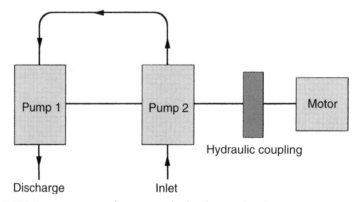

**Figure 4-55** Pumps connected in series—hydraulic coupling between pump 2 and motor.

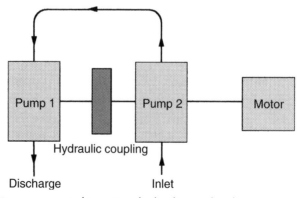

**Figure 4-56** Pumps connected in series—hydraulic coupling between pump1 and pump 2.

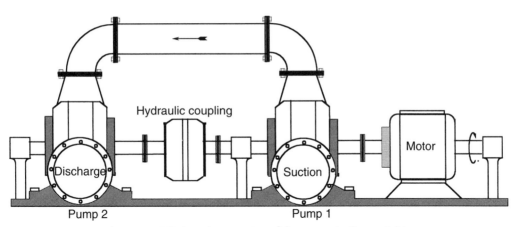

**Figure 4-57** Actual connection of the pumps in Figure 4-56.

### 4.7.1 Multistage Centrifugal Pumps (see Figures 4-58 and 4-59)

Other examples of multistaged pumps are those used for pumping wells or deep boreholes (see Figures 4-60 and 4-61). These fall into two categories:

1. Shaft-driven pumps
2. Completely submersible pumps

# 4.8 Displacement Rotary Pumps

### 4.8.1 Vane Pumps

A schematic diagram of a typical vane pump is shown in Figure 4-62. The pump operates as follows. After inlet, liquid that is contained and sealed between neighboring chambers is translated to the outlet port by clockwise rotation of the chamber. Inlet and outlet ports are slots in the sides as shown. Sealing of the vanes on the inside periphery of the casing is maintained either by the action of centrifugal forces on the vanes or by springs pushing the vanes outwards to the casing. In this way, the system self-compensates for wear. The nature of the design is such that vane pumps do not lend themselves to the generation and maintenance of high pressures. Usually, their pressure range is up to 20 atmospheres. Their use lies in the ability to pump liquids at large volumetric flow rates. Consequently, they have higher volumetric efficiencies than other pumps.

A pump similar to that shown in Figure 4-62 but with flexible vanes is shown in Figure 4-63. The mode of operation is similar to the previous pump, except that fluid is in the chambers contained by adjacent flexed vanes. A typical operating envelope for such pumps is shown in

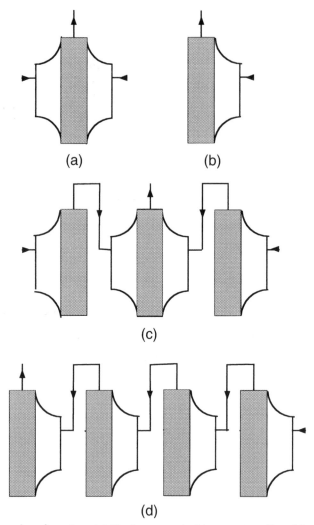

**Figure 4-58** Examples of staging. (a) Single-stage, double-entry impeller; (b) single-stage, single-entry impeller; (c) single-stage, double-entry impeller connected to two single-stage, single-entry impellers; (d) four single-stage, single-entry impellers connected in series.

Figure 4-64. There is a linear relationship between H and Q for any pump encompassed by this figure; for example, the chained-dotted lines represent characteristics at each end of the range.

## 4.8.2 Peristaltic Pump

Peristaltic pumps are positive displacement pumps in which only the tube or flexible pipe touches the pumped fluid. Fluid is drawn into the pump and is trapped between two rollers or shoes.

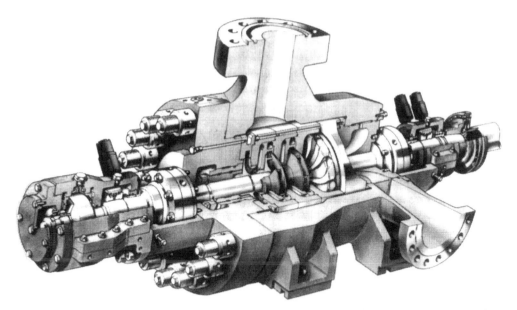

**Figure 4-59** Cutaway diagram of a multistage, radial impeller, centrifugal pump. (Courtesy Sulzer Pumps Ltd.)

As shown in Figure 4-65, as the top shoe rotates clockwise, it expels fluid in front of it. As the tube is released causing it to spring back to its original position, the bottom shoe starts to squeeze the tuning. This action gives a positive displacement to the fluid. In this way there is always positive closure, and backflow is prevented. Usually, such pumps are used for small flow rates and are easily adapted to make accurate metering of fluid possible. However, large peristaltic pumps can handle several thousand liters per minute at pressures of three atmospheres and above.

Such a pump has a number of advantages over other pumps for small flow rates where fluids must not be contaminated or changed. Examples are in the pharmaceutical and food industries and in medicine where extra corporeal pumping is needed.

### 4.8.3 Lobe Pumps (see Figures 4-66 and 4-67)

### 4.8.4 RVP Pump

The RVP pump (Round et al. 1997) is a rotary displacement pump with a series of intermeshing lobes around two connected rotors eccentrically positioned on the same shaft. The pump offers a number of advantages over similar pumps in that the volumetric efficiency approaches that of vane pumps while providing a high delivery pressure. Since the rotors are rotating in the same direction, the relative velocity between them is small, thus ensuring small internal leakage.

**Figure 4-60** Shaft-driven borehole pump.

The gaps between the rotors are labyrinthine, further enhancing low gap loss. (See Figures 4-68 and 4-69.)

### 4.8.5 Water Ring Pumps

Water ring pumps are fixed vane-type pumps in which the impellers are eccentrically located (see Figure 4-70). The water enclosed between vanes acts in a manner similar to the plunger

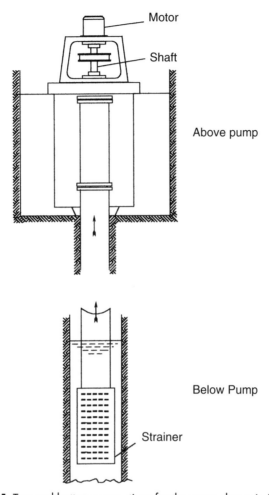

**Figure 4-61** Top and bottom connections for the pump shown in Figure 4-60.

of a reciprocating pump. During the first half rotation, air is sucked in, compressed, and pushed out.

## 4.9 Flow Control

Flow control of pumps has much in common with flow control of turbines. There are several methods of control, each of which has its own advantages and disadvantages: throttling of the flow at inlet or outlet; disconnection of one or more pumps of a multiconnected pump system; regulated flow bypass; speed regulation; impeller blade adjustment; inlet guide-vane adjustment; and air locking.

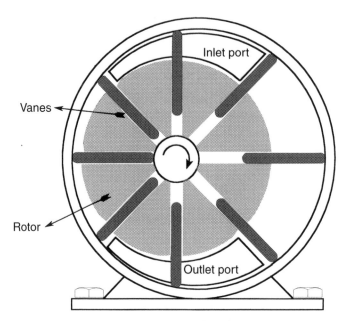

**Figure 4-62** Vane pump with eight rigid vanes.

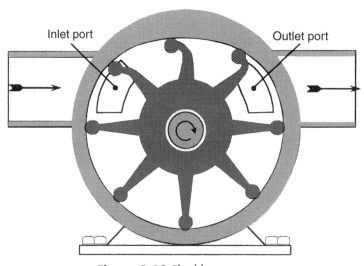

**Figure 4-63** Flexible vane pump.

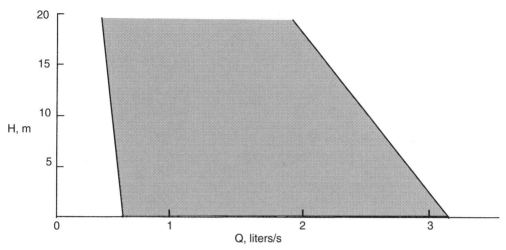

**Figure 4-64** Typical operating envelope for flexible vane pumps.

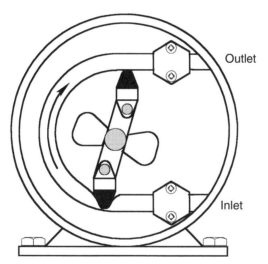

**Figure 4-65** A peristaltic pump showing the mode of action.

Cavitation management is also used on small pumps, but generally speaking cavitation is to be avoided and flow control by this method is not considered viable.

## 4.9.1 Throttling of the Flow at Inlet or Outlet

This method of flow control should be used only where flow rate changes are required for short periods of time. The connecting system has an increased resistance and intersects the pump characteristic at a lower operating point. At low flow rates, pumps run unevenly. Performance

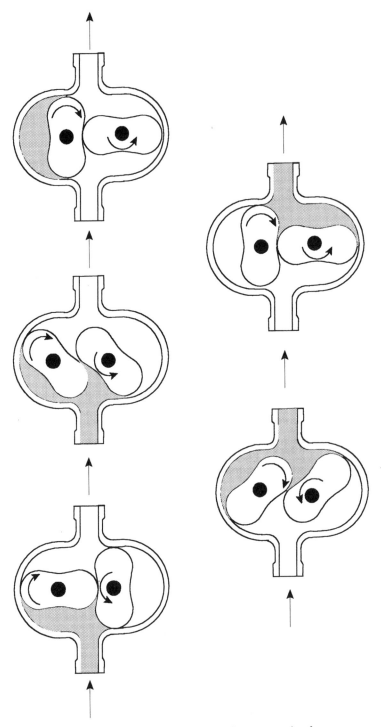

**Figure 4-66** Double-lobe, two-axis pump showing mode of pumping.

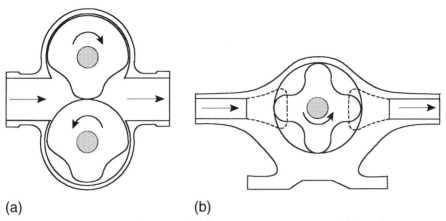

(a)                                    (b)

**Figure 4-67** Two other types of lobe pump: (a) two-axis; and (b) single-axis.

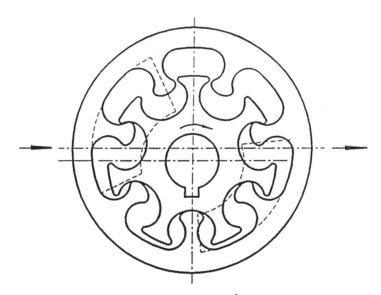

**Figure 4-68** Cross section of RVP pump.

is somewhat better at low flow rates for a pump with a flat characteristic; that is, this suggests a radial impeller with a lower value of $N_S$.

## 4.9.2 Pump Disconnection

For flow rates that vary widely, it is often more useful for better control to have several pumps connected in parallel. When one or more of the pumps are stopped, the remaining ones operate

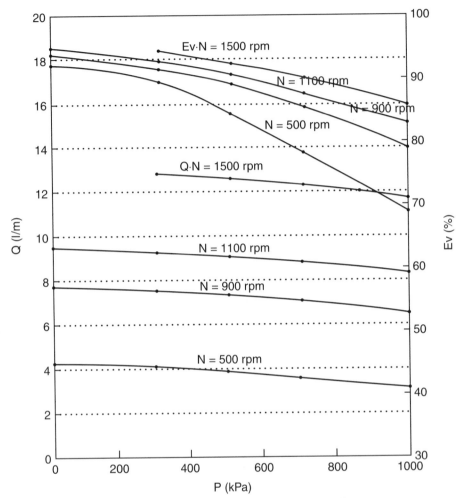

**Figure 4-69** Experimental data from tests on a 6 cm outside diameter RVP pump.

closer to their most efficient condition. Series connections usually should not be used in this way because the remaining pump or pumps must move to the right on the characteristic curve, which means that the NPSH available cannot cover the NPSH required. In addition, series operation puts more stress on seals and casings.

## 4.9.3 Regulated Flow Bypass

A regulated bypass flow controller means that part of the main stream is diverted back to the suction side of the pump (see Figure 4-71). Because of discharge loss in the bypass line, the overall efficiency is reduced. It is therefore the least satisfactory method of flow control for

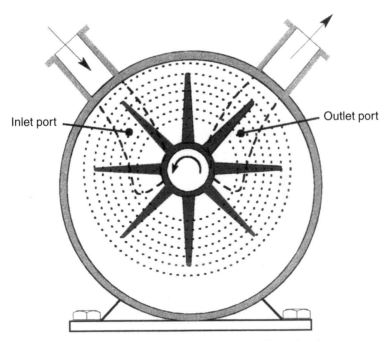

Inlet port

Outlet port

**Figure 4-70** A water ring pump, showing inlet and outlet ports.

large-capacity centrifugal pumps. It is often used for flow control at the research laboratory level where efficiency is not of prime importance or for fairly small pumps. Figure 4-72 shows a small boiler feed pump where bypassing is used for pump balancing. If used on a larger scale commercially, an axial-flow type pump would be more suitable because of the steep characteristic at low flow rates and where power fall as the flow rate increases.

## 4.9.4 Speed Regulation

Speed control of the pump is much to be preferred over other methods, although it is more expensive. Transition for different flow rates is smooth, especially for system heads that derive mostly from friction losses. Throttling causes the system characteristic to change, but with speed control the pump characteristic changes. Usually, the efficiency remains fairly constant over a practical speed change range. An added advantage is that the life of the system is increased.

Speed control is achieved by connecting the pump in the following ways:

1. Variable-speed prime movers such as diesel engines and variable-speed electric motors
2. Electromagnetic, hydrostatic, or hydraulic couplings
3. Variable-speed gears

The affinity laws are applicable to speed changes (see Chapter 2, Section 2.8).

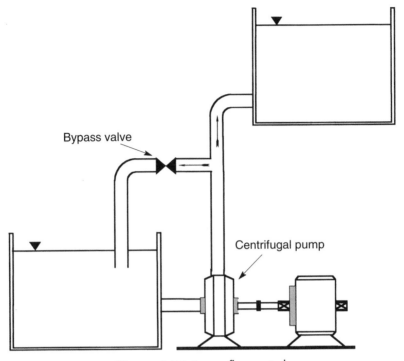

**Figure 4-71** Bypass flow control.

### 4.9.5 Impeller Blade Adjustment

Impeller blade adjustment occurs by means of variable pitch blades—similar to variable blades on a Kaplan turbine. The adjustment of the blades is made at the pump hub. This translates into changing the pump characteristic at constant pump speed for each blade position. However, at low flow rates the characteristics of mixed-flow and axial-flow pumps are unstable because of high blade loading.

At high head, the possibility of cavitation is increased together with the NPSH. Water circulation pumps make good use of this method for constant water level systems.

### 4.9.6 Inlet Guide-vane Adjustment

In this case, the impeller blades do not move and are attached rigidly to the hub. The end effect is similar to impeller blade adjustment in that the pump characteristic is changed. A tangential velocity component is imparted to the inlet fluid, changing the energy conversion and thus the pump characteristic.

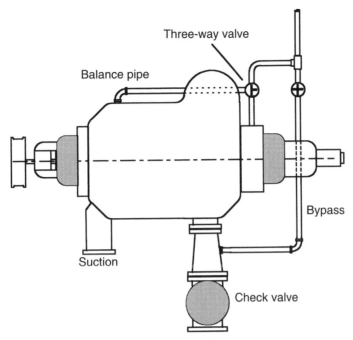

**Figure 4-72** Bypass flow and balance control of a small centrifugal pump.

### 4.9.7 Air Locking

The physical arrangement of pump-pipe systems is important so that air locking or air bubble formation is avoided. This usually involves simple realigning of the pipe and fittings so that this does not occur. Figure 4-73 shows three such arrangements that will cause bubble formation, and the shaded areas indicate where air bubbles will form. In each inlet pipe section of Figures 4-73(a), (b), and (c), the liquid will still flow, but each system's overall efficiency will be decreased. However, if each condition is exaggerated so that the bubble becomes much larger, then the system will be greatly reduced or flow will entirely cease and the system will be completely air-locked. The resolution of these problems is simple once the difficulty and location of the air lock are established.

## 4.10 Automatic Priming

Priming of pumps may be categorized according to the size of the pump. For small pumps, that is, 1.5 to 110 m head and up to 0.13 m$^3$/s, if priming is required, the priming system usually becomes an integral part of the pump.

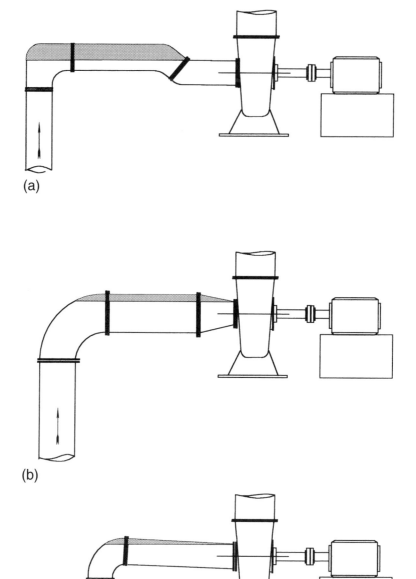

(a)

(b)

(c)

**Figure 4-73** Three pump-pipe systems that lead to flow blockage by air locking or bubble formation: (a) realignment of pipe needed; (b) uniform pipe diameter needed; (c) another form of pipe misalignment.

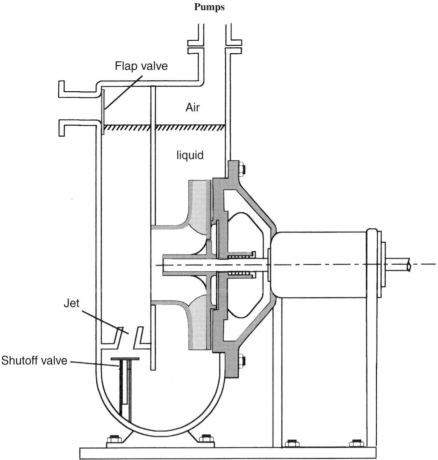

**Figure 4-74** Nonrunning condition of self-priming small pump showing priming jet with shutoff valve open.

Designs for smaller pumps fall into two classes:

1. Pumps with modified impellers such as chevron-bladed impellers and those that can act as dual service pumps (i.e., capable of pumping air and liquid); an example is a liquid ring pump such as that illustrated in Figure 4-70.
2. Pump systems fitted with priming nozzles.

Larger pumps have external priming arrangements. Priming for a small pump that is an integral part of the pump is illustrated in Figures 4-74 and 4-75. It will be immediately noticed that the pump casing is of a different form from a normal volute casing. In Figure 4-74 the pump is shown in its stopped or nonrunning condition (i.e., the flapper valve is closed, and the shutoff valve is open). In Figure 4-75 the pump is shown in its running condition (i.e., the flapper valve is open, and the shutoff valve is closed by virtue of the pressure of the exiting fluid acting underneath

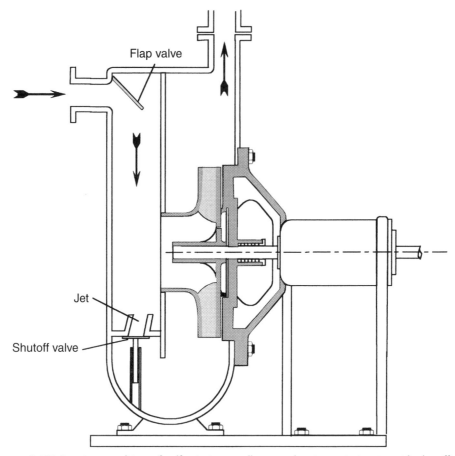

**Figure 4-75** Running condition of self-priming small pump showing priming jet with shutoff valve closed.

the shutoff valve). In the nonrunning condition, liquid is shown level with the inlet pipe, and the shutoff valve has fallen to the bottom of its travel by virtue of its weight.

When the pump is started, the impeller sucks in the fluid immediately opposite. A partial vacuum results in the intake chamber, causing the liquid level to fall and air to be taken into the suction side of the pump. At this point, a mixture of liquid and air bubbles flows to the suction side of the impeller. The shutoff valve begins to rise, but while it is still open liquid is jetted through the nozzle directly to the suction side of the impeller. This continues until sufficient pressure is built up underneath the valve until it closes. At this point only liquid is pumped. This is illustrated in Figure 4-75.

An alternative jet nozzle arrangement is shown in Figure 4-76. In this system, the priming device is fitted onto the suction and discharge pipes, with an air separator connecting the two.

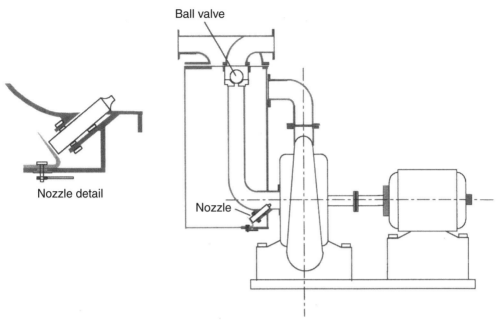

**Figure 4-76** Priming system for a self-priming small pump with separated tank from the pump.

One advantage of this arrangement over the system in Figure 4-74 is that the pump itself is a conventional one that is self-contained, so that the volute is a conventional one. Once priming has been completed, the nozzle may be closed manually or automatically, with a shutoff fitted to the separator tank. The priming device may be added when the pump is built or retrofitted. A further advantage is that the priming time is much reduced. A self-priming system for a larger pump than that shown in Figure 4-76 is shown in Figure 4-77. Figures 4-78 and 4-79 show two external arrangements for large pumping systems.

## 4.11 Fluid Couplings

Fluid couplings are used to transmit power from one shaft to another in a smooth, controlled way. They may be divided as follows:

1. Hydrokinetic
2. Hydrodynamic
3. Hydroviscous
4. Hydrostatic

Hydrokinetic couplings are more commonly known as hydraulic couplings. They consist of a driving member, the impeller, and a driven member, the runner. Figure 4-80 illustrates such

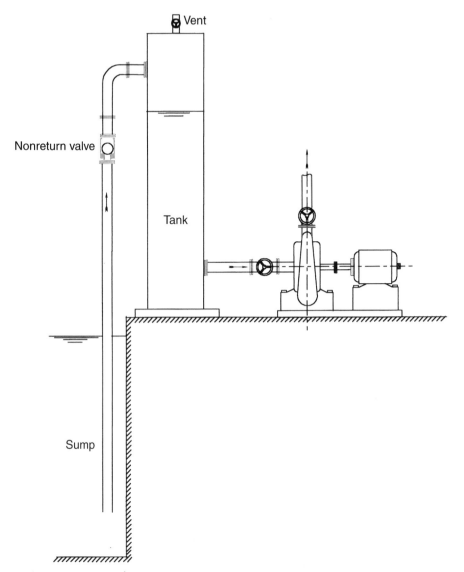

**Figure 4-77** Simple priming system using a priming tank for a medium-size pump.

a coupling schematically. The working fluid is always oil, either natural or synthetic, because the oil serves as a lubricant and a coolant.

The available power to be delivered from one shaft to the other is equal to the difference in kinetic energy of the fluid coming from the impeller and the kinetic energy of the fluid coming from the runner. If the mass of fluid flowing through the device is constant, then the resulting speed

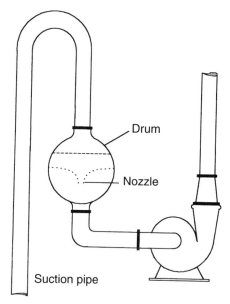

Figure 4-78 Automatic priming using a combined drum and nozzle.

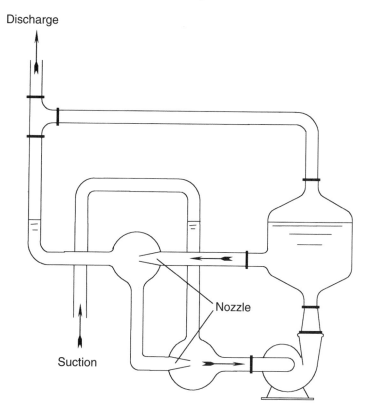

**Figure 4-79** Automatic priming using a separation chamber and two nozzles.

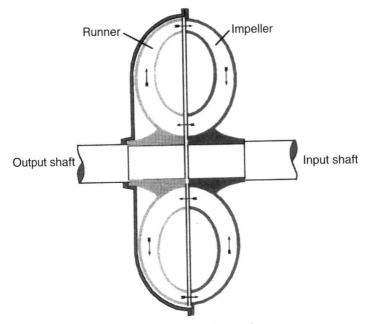

**Figure 4-80** Hydraulic coupling.

is constant. Föttinger (1910) was a pioneer in developing this type of transmission or coupling, which has been variously known as a Föttinger transmission or Vulkan coupling. Vulkan couplings usually refer to a constant torque device. Their efficiency is high; where the torques of the driver and driven shafts are the same, efficiencies can reach 99%.

If the mass is varied either by bypassing with a scoop or by leakage through the outer casing through ports, then the speed may be varied. Because a portion of the theoretical energy in the fluid is lost though thermodynamic inefficiency, and in order to keep the working fluid from overheating as a result of the heat that is released into the fluid, the coupling is surrounded by a chamber containing an oil sump. The oil may be pumped externally through a cooling system, and the cooled fluid may be injected into the coupling. The chamber further serves as a control system for vapor or liquid loss. This variant on the transmission has a lower efficiency than the constant-speed device. Figure 4-81 illustrates such a system. It will be noticed that there are leak ports for the oil on the outside casing and an inlet port for cooled incoming oil on the inner casing.

Originally, the Föttinger transmission was intended for ship propulsion where it was necessary to transmit power from a high-speed turbine to a low-speed ship propeller. Later, the transmission found successful application in land vehicles: locomotives and automobiles. Figure 4-81 shows the hydraulic coupling of Figure 4-80 fitted with a cooling system.

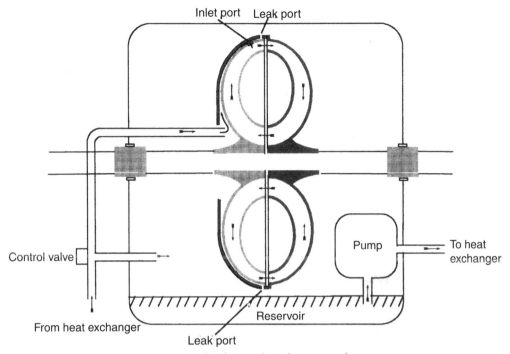

**Figure 4-81** Hydraulic coupling showing cooling system.

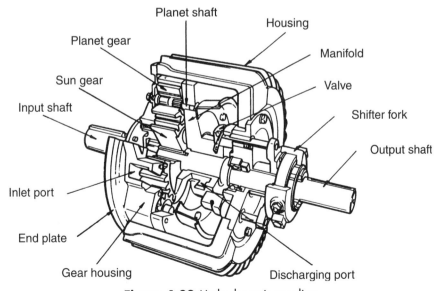

**Figure 4-82** Hydrodynamic coupling.

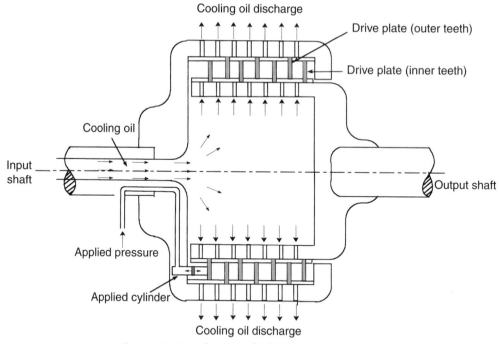

**Figure 4-83** Schematic of a hydroviscous coupling.

Hydrodynamic couplings have planetary gear drives. They are not common, and when they are used, they are for portable pumps. The components consist of an input shaft driving the housing, an end plate, and planetary gear shafts. The planetary gears are partially surrounded by the manifold that forms a cavity. A sun gear drives the output shaft. An example of such a coupling is illustrated in Figure 4-82.

Hydroviscous couplings depend on the variation of viscous shearing forces in oil contained between rotating parallel plates. Varying the separation distance of the plates changes the shearing forces. See Figure 4-83.

Hydrostatic couplings use a positive displacement hydraulic pump coupled to a positive displacement motor. In many cases, fluid is bypassed from delivery back to suction. This provides a continuously variable flow and consequently a variable output speed. Another variant uses a variable-flow positive displacement pump; this in turn can vary output speed.

## 4.12 Solved Problems

### 4.12.1 Homologous Pumps

The experimental data shown in Table 4-1 are available for a centrifugal pump operating at 3550 rpm with a 9-cm-diameter impeller.

**Table 4-1**

| H, m | Q × 10⁵ m³/s | η % |
|------|--------------|-----|
| 5.35 | 0 | 0 |
| 5.31 | 3.6 | 14 |
| 5.27 | 7.2 | 26 |
| 5.22 | 10.7 | 39 |
| 5.17 | 14.3 | 51 |
| 5.04 | 28.4 | 63 |
| 4.83 | 35.5 | 73 |
| 4.54 | 42.6 | 79 |
| 4.1 | 49.7 | 82 |
| 3.59 | 56.7 | 81 |
| 2.95 | 63.8 | 77 |
| 2.26 | 70.9 | 72 |

What size homologous pump is required to produce $Q = 0.00592$ m³/s at the best efficiency with a head of 23.9 m? Plot the characteristics of the new pump.

### Solution

An examination of the experimental data shows that the best efficiency is 82%. At this efficiency $H_1 = 4.26$ m and $Q_1 = 49.7 \times 10^{-5}$ m³/s.

The appropriate dimensionless groups are:

$$(H_1/N_1^2 D_1^2) = (H_2/N_2^2 D_2^2) \tag{4.54}$$

and

$$(Q_1/N_1 D_1^3) = (Q_2/N_2 D_2^3) \tag{4.55}$$

Substituting the appropriate values:

$$4.1/(3500^2 \times 9^2) = 23.9/(N_2^2 D_2^2) \tag{4.56}$$

$$49.7 \times 10^{-5}/(3500 \times 9^3) = 0.00592/(N_2 D_2^3)$$

Solving for $N_2$ and $D_2$: $N_2 = 3800$ rpm and $D_2 = 20$ cm.

The equations for transforming the corresponding values of H and Q are:

$$H_2 = H_1(N_2^2 D_2^2)/(N_1^2 D_1^2) = 5.82 \quad \text{and} \tag{4.57}$$

$$Q_2 = Q_1(N_2/N_1)/(D_2^3/D_1^3) = 11.91 \, Q_1 \tag{4.58}$$

The new data are shown in Table 4-2 and plotted in Figure 4-84.

**Table 4-2**

| H, m | Q × 10⁵ m³/s | η % |
|------|--------------|------|
| 31.1 | 0 | 0 |
| 30.9 | 42.9 | 14 |
| 30.7 | 85.8 | 26 |
| 30.4 | 127.4 | 34 |
| 30.1 | 170.3 | 43 |
| 29.3 | 338.2 | 69 |
| 28.1 | 442.8 | 76 |
| 26.4 | 507.4 | 79 |
| 23.9 | 591.9 | 88 |
| 20.9 | 675.3 | 80 |
| 17.2 | 759.9 | 76 |
| 13.2 | 844.4 | 72 |

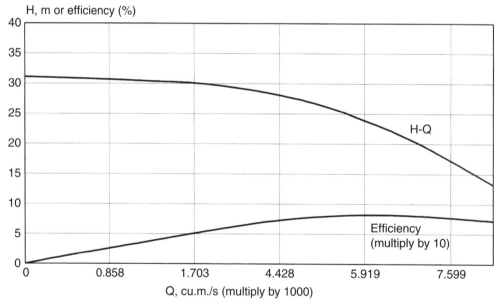

**Figure 4-84** Characteristics of the new pump.

## 4.12.2 Use of Slip Factor

A centrifugal pump has a $Q = 0.1$ m³/s and $N = 1200$ rpm. The impeller has seven backward-leaning blades with a blade outlet angle $\beta_2 = 50°$. Impeller OD = 0.4 m: ID = 0.2 m and an impeller width at exit of 31.7 mm. Assuming a diffuser efficiency = 51.5%: impeller head losses = 10% of the theoretical head rise and the diffuser exit is 0.15 m in diameter, what are:

1. Slip factor?
2. Manometric head?
3. Hydraulic efficiency?

## Solution

The Busemann slip factor equation will be used.

The criterion for constant evaluation is:

$$\exp(2\pi \cos \beta_2/z) = \exp(2\pi\, 0.643/7) = 1.78 \tag{4.59}$$

$(r_2/r_1) = 2$, which is $> 1.78$

Therefore, B may be taken to be $= 1$, and A, interpolated from Figure 4-17, is approx. $= 0.75$.

$$\text{Blade tip speed } u_2 = \pi N D_2/60 = (\pi)(1200)(0.4)/60 = 25.13 \text{ m/s}$$

$$\text{Radial velocity } c_{r2} = Q/(\pi D_2 b_2) = 2.51 \text{ m/s}$$

The Busemann slip factor equation is:

$$\sigma_B = (A - B\Phi_2 \tan \beta_2)/(1 - \Phi_2 \tan \beta_2) \tag{4.60}$$

Substituting values: $\sigma_B = 0.74$.

Hydraulic losses occur in the impeller and the diffuser. Kinetic energy leaving the diffuser is not or only partially recovered and contributes to the total loss. The loss of head in the diffuser is:

$$H_D = (1 - \eta_D)(c_2^2 - c_3^2)/2g \tag{4.61}$$

Other losses are: Impeller $= (0.1)(u_2 c_{u2})/g$; exit $= c_3^2/2g$

Summing all the losses:

$$H_L = H_D + H_{IMP} + H_E = (0.485)(c_2^2 - c_3^2)/2g + (0.1)(0.739)(u_2 c_{u2})/g + c_3^2/2g \tag{4.62}$$

$$c_{u2} = \sigma_{Bu_2}(1 - \Phi_2 \tan \beta_2^1)$$

Substituting values: $c_{u2} = 16.5$ m/s

Substituting values for $H_I$:

$$H_I = 41.8\,\text{m}: c_2^2/2g = 14.0\,\text{m}: c_3 = (4Q)/(\pi d_{DIFF}^2) = 5.7\,\text{m}: c_3^2/2g = 1.7\,\text{m} \qquad (4.63)$$

Substituting values for $H_L$: $H_L = 11.8\,\text{m}$

The manometric head is: $H = H_I - H_L = 30.0\,\text{m}$

The hydraulic efficiency: $\eta_{HYD} = H/H_I = 71.7\%$

### 4.12.3 Pressure Delivered and Energy Required for a Centrifugal Pump

A radial-flow pump operates at steady state with constant rotational speed of 900 rpm. The impeller eye radius is 5 cm, and the outside diameter is 40 cm. The vane height is constant at 6.4 cm, and vane angles: $\beta_1 = 75°$; $\beta_2 = 83°$. The overall efficiency of the pump is 89%. Find:

1. The volumetric flow rate
2. The rise in stagnation pressure and the increase in static pressure across the impeller
3. Power transferred to the fluid
4. Input shaft power

### Solution
The inner tip velocity is:

$$U_1 = \omega R_1 = (900H\,2\pi/60)(0.05) = 4.7\ \text{m/s} \qquad (4.64)$$

$$(U_1/V_{r1}) = \tan\beta_1 \qquad (4.65)$$

$$V_{r1} = 4.71/\tan 75° = 1.3\ \text{m/s}$$

Volumetric flow rate

$$Q = A_1 V_{r1} = 2\pi R_1 b V_{r1} = 2\pi(0.05)(0.064)(1.26) = 0.025\ \text{m}^3/\text{s} \qquad (4.66)$$

The continuity equation may be written:

$$R_1 V_{r1} = R_2 V_{r2} \qquad (4.67)$$

$$V_{r2} = (0.05)(1.26)/(0.20) = 0.32\ \text{m/s}$$

The outer tip velocity is:

$$U_2 = \omega R_2 = (900 \times 2\pi/60)(0.20) = 18.9\ \text{m/s} \qquad (4.68)$$

The stagnation pressure rise across the impeller is given by:

$$p_{02} - p_{01} = \rho\, U_2^2\, \eta\, [1 - (V_{r2}/U_2)\tan \beta_2]$$

$$= (1000)(18.9)^2(0.89)[1 - (0.32/18.9)(8.14)]/1000 = 272\,\text{kPa}$$

$$V_2 = U_2 - V_{r2} \tan \beta_2 = 18.9 - (0.32)(8.14) = 16.3\,\text{m/s} \qquad (4.69)$$

$$\tan \alpha_2 = V_2/V_{r2} = 16.3/0.32 = 50.8: \alpha_2 = 88.9°$$

From the definition of static pressure:

$$p_2 - p_1 = p_{02} - p_{01} + (\rho/2)(V_1^2 - V_2^2)$$

$$= [272,000 + 1000/2(1.59 - 264.06)]/1000 = 140.8\,\text{kPa} \qquad (4.70)$$

Energy received by the fluid:

$$(dW/dt) = Q(p_{02} - p_{01}) = (0.0253)(272,000)/1000 = 6.9\,\text{kW} \qquad (4.71)$$

Input power to shaft: $(dW/dt)_S = 6.9/0.89 = 7.75\,\text{kW}$ (4.72)

### 4.12.4 Pressure Rise through an Impeller with a Diffuser

A centrifugal pump has backward-leaning vanes of outlet angle $\beta_2$. The flow velocity through the impeller is constant, that is, $c_{r1} = c_{r2} = c_r$, and the absolute inlet velocity is radial. The pump is fitted with a diffuser. Assuming that the flow is frictionless with no circulatory or separation losses, show that the pressure change across the impeller divided by the work done per unit-specific weight of fluid flowing for a pump *without a diffuser* is:

$$(2)[1 + (c_r \cot \beta_2/u_2)] \qquad (4.73)$$

and *with a diffuser* fitted:

$$(2)[(1 + k) + (1 - k)(c_r \cot \beta_2/u_2)] \qquad (4.74)$$

where:

k = fraction of whirl component converted to pressure energy by the diffuser.

### Solution

Referring to Figure 4-8: $c_{r2} = (u_2 - c_{u2}) \tan \beta_2.$

Rewriting: $c_{u2} = (u_2 - c_{r2}) \cot \beta_2$

The work done/unit-specific weight of fluid: $= (c_{u2}u_2)/g = (u_2/g)(u_2 - c_{r2}) \cot \beta_2$ (4.75)

The work done/unit-specific weight of fluid for no losses $= H_2 - H_1$

First, applying the Bernoulli equation between the inside (1) and the outside (2) of the impeller with *no diffuser* present:

$$(p_2\gamma) - (p_1/\gamma) = c_1^2/2g - c_2^2/2g + (u_2/g)(u_2 - c_{r2})\cot\beta_2 \qquad (4.76)$$

$$c_1 = c_{r1} = c_{r2} = c_r$$

Thus:

$$c_1^2/2g = c_r^2/2g$$

$$c_2^2 = c_r^2 + c_{u2}^2 = c_r^2 + [(u_2 - c_{r2})\cot\beta_2]^2 \qquad (4.77)$$

Rearranging:

$$(p_2 - p_1)/\gamma = (1/2g)(u_2^2 - c_{r2}^2\cot^2\beta_2)$$

$$\therefore 4[(p_2 - p_1)/\gamma]/(\text{workdone/unit-specific weight}) = (2)[1 + (c_r\cot\beta_2/u_2)] \qquad (4.78)$$

With a diffuser fitted:

$$(p_2 - p_1)/\gamma = (\text{work done/unit specific weight}) - V_2^2/2g + c_{r2}^2/2g + kc_{u2}^2/2g \qquad (4.79)$$

$$= (u_2/g)(u_2 - c_{r2})\cot\beta_2 - c_{r2}^2/2g - [(u_2 - c_{r2})\cot\beta_2]^2/2g + c_{r2}^2/2g$$

$$+ [(k/2g)(u_2 - c_{r2})\cot\beta_2]^2 \qquad (4.80)$$

Dividing both sides by (work done/unit-specific weight), that is, $(u_2/g)(u_2 - c_{r2})\cot\beta_2$ and rearranging:

$$[(p_2 - p_1)/\gamma]/(\text{work done/unit specific weight}) = (1/2u_2)[u_2(1 + k) + (1 - k)c_{r2}\cot\beta_2] \qquad (4.81)$$

or

$$[(p_2 - p_1)/\gamma]/(\text{work done/unit specific weight}) = (2)\{(1 + k) + [(1 - k)c_r\cot\beta_2/u_2]\} \qquad (4.82)$$

### 4.12.5 Identical Centrifugal Pumps Connected in Series and Parallel

Tests on a centrifugal pump produced the H-Q and efficiency data shown in Figure 4-85. Two such pumps are connected to run in series and parallel. The external resistance against which the pumps are to work is represented by the system curve in the same figure.

Determine:

1. The discharge when two pumps are working in parallel
2. The discharge when two pumps are working in series
3. The power required for conditions 1 and 2

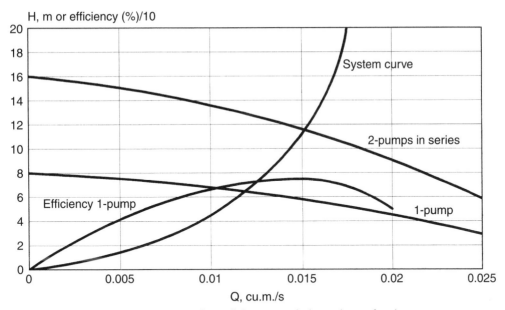

**Figure 4-85** Series- and parallel-connected identical centrifugal pumps.

## Solution

1. The value of Q for a parallel connection is twice the Q for a single pump at the same head. The point of intersection of the H-Q curve and the system curve is at: $H = 7.3$ m and $Q = 0.0125$ m$^3$/s at an efficiency of 74%.
   The power required is:

$$\gamma HQ/\eta = (1000)(9.81)(7.3)(0.0125)/(0.74)$$
$$= 1210 \text{ watts} = 1.21 \text{ kW} \tag{4.83}$$

2. The point of intersection of the H-Q curve and the system curve for series connection is:

$$H = 11.6 \text{ m} \quad \text{and} \quad Q = 0.0153 \text{ m}^3/\text{s at an efficiency of 76\%.} \tag{4.84}$$

   The power required is:

$$\gamma HQ/\eta = (1000)(9.81)(11.6)(0.0153)/(0.76)$$
$$= 2291 \text{ watts} = 2.29 \text{ kW} \tag{4.85}$$

## Comment

This problem illustrates the importance of the shapes of the characteristic curves of individual pumps on the final combination curve. Because of the relative flatness of the basic H-Q curve,

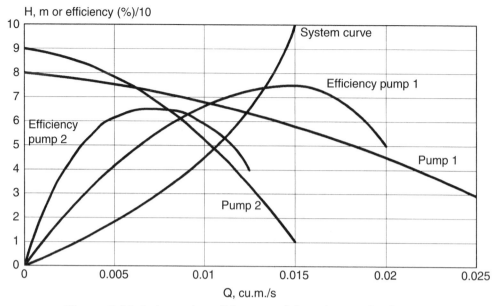

**Figure 4-86** Series- and parallel-connected dissimilar centrifugal pumps.

a parallel combination of pumps shows little improvement over a single pump. The operating point of the combination is quite close to the operating point of the single pump. So there is no real advantage to be obtained here. The series combination, on the other hand, shifts the operating point much further to the right, resulting in a higher head and volumetric flow rate. There is also a slight improvement in efficiency.

## 4.12.6 Nonidentical Centrifugal Pumps Connected in Series and Parallel

Two centrifugal pumps whose characteristics are shown in Figure 4-86 are connected both in series and parallel. The system curve is also shown. What is the head developed, the total discharge, and power requirement for each combination for the given system curve?

### Solution

The experimental data for each centrifugal pump are shown in Figure 4-86. The pumps are connected to run in series and parallel. The external resistance against which the pumps are to work is represented by the system curve in the same figure.

   1. Parallel connection—in this case the $Q = s$ are additive.

     From Figure 4-86 at the operating point, $H = 7.4$ m and $Q = 0.0133$ m$^3$/s.

     The power required by pump (1) is: $P_1 = \gamma_1 H_1 Q_1 / \eta_1$.

The value of $\eta_1$ at the operating point is 73%, and the head is 7.3 m.

The value of $Q_1 = 0.0070 \text{ m}^3/\text{s}$.

Therefore:

$$P_1 = (1000)(9.81)(7.3)(0.007)/(0.73)$$

$$= 687 \text{ watts} = 0.687 \text{ kW} \tag{4.86}$$

The power required by pump (2) is: $P_2 = \gamma_2 H_2 Q_2 / \eta_2$.

The value of $\eta_2$ at the operating point is 27%, and the head is 7.3 m.

The value of $Q_2 = 0.0063 \text{ m}^3/\text{s}$.

Therefore:

$$P_2 = (1000)(9.81)(7.3)(0.0063)/(0.27)$$

$$= 1671 \text{ watts} = 1.671 \text{ kW} \tag{4.87}$$

The total power required for this combination is therefore: $0.687 + 1.671 = 2.358 \text{ kW}$.

2. Series connection—the heads are additive such that the value of the sum falls on the system curve. In this case the sum is $H = 8.2$ m, giving a $Q = 0.014 \text{ m}^3/\text{s}$.

The power required by pump (1) is: $P_1 = \gamma_1 H_1 Q_1 / \eta_1$.

The value of $\eta_1$ at the operating point is 75%, and the head is 6.0 m.

Therefore:

$$P_1 = (1000)(9.81)(6.0)(0.014)/(0.75)$$

$$= 1098 \text{ watts} = 1.098 \text{ kW} \tag{4.88}$$

The power required by pump (2) is: $P_2 = \gamma_2 H_2 Q_2 / \eta_2$.

The value of $\eta_2$ at the operating point is 15%, and the head is 2.2 m.

The value of $Q_2 = 0.014 \text{ m}^3/\text{s}$.

Therefore:

$$P_2 = (1000)(9.81)(2.2)(0.014)/(0.15)$$

$$= 2014 \text{ watts} = 2.014 \text{ kW} \tag{4.89}$$

The total power required for this combination is therefore: $1.098 + 2.014 = 3.112 \text{ kW}$.

## Comment

The better combination in this case would be the parallel one. The values of Q are approximately the same in each case, and H differs by about 10%. However, the power is different by over 50%.

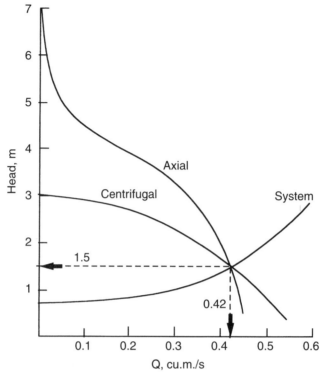

**Figure 4-87** Starting conditions for an axial and centrifugal pump connected to the same system.

### 4.12.7 Comparison of Starting Conditions of an Axial and a Centrifugal Pump

An axial-flow and a centrifugal pump are connected in sequence to the same pipe, 1000 m long and 60 cm in diameter, containing water. The characteristics of each pump are shown in Figure 4-87. Compare the times taken to reach the operating condition starting from t = 0 and the power absorbed in the process. Ignore pressure waves and consider only the mean acceleration of the water in the pipe.

### Solution

The head accelerating the fluid for each pump is the difference between that for each pump and the system curve. The x-axis is divided into a number of finite intervals; for accuracy and convenience, in the present case $\Delta Q = 0.02$ m³/s was chosen. Thus, there are 21 intervals. The mean acceleration for each interval is given by:

$$a\Delta t = 0.02/[(\pi/4)(0.60)^2] \qquad (4.90)$$

If the height of the pump curve above the system curve is designated as $h_i$, then the mean acceleration is also:

$$a = (h_i g)/1000 \qquad (4.91)$$

Combining both equations:

$$\Delta t = 7.21/h_I \qquad (4.92)$$

Using this equation, the following times were calculated at the beginning and at the end of each interval.

For the axial pump:

$\Delta t$, s

| | | | | | | | | | | |
|---|---|---|---|---|---|---|---|---|---|---|
| 1.03 | 1.25 | 1.31 | 1.49 | 1.57 | 1.60 | 1.64 | 1.72 | 1.76 | 1.80 | 1.83 |
| 1.90 | 1.92 | 2.00 | 2.10 | 2.22 | 2.33 | 2.49 | 2.62 | 3.00 | 3.61 | 4.80 |

$\Delta t$ (av) s

| | | | | | | | | | | |
|---|---|---|---|---|---|---|---|---|---|---|
| 1.14 | 1.32 | 1.44 | 1.53 | 1.59 | 1.62 | 1.68 | 1.74 | 1.78 | 1.82 | 1.87 |
| 1.91 | 1.96 | 2.05 | 2.16 | 2.28 | 2.41 | 2.56 | 2.81 | 3.31 | 4.21 | |

$\Sigma = 43.2$ s

Power absorbed per interval:

$$\Delta P = (h_i)\Delta q \rho g = 1414/\Delta t \qquad (4.93)$$

$\Delta P$, kW

| | | | | | | | | | | |
|---|---|---|---|---|---|---|---|---|---|---|
| 1.24 | 1.07 | 0.98 | 0.92 | 0.89 | 0.87 | 0.84 | 0.81 | 0.79 | 0.78 | 0.77 |
| 0.74 | 0.72 | 0.69 | 0.65 | 0.62 | 0.59 | 0.55 | 0.50 | 0.43 | 0.34 | |

Total $= 15.79$ kW

For the centrifugal pump:

$\Delta t$, s

| | | | | | | | | | | |
|---|---|---|---|---|---|---|---|---|---|---|
| 2.41 | 2.44 | 2.49 | 2.51 | 2.53 | 2.56 | 2.57 | 2.58 | 2.62 | 2.67 | 2.74 |
| 2.83 | 2.88 | 3.00 | 3.20 | 3.35 | 3.52 | 3.70 | 3.90 | 4.01 | 4.12 | 4.80 |

$\Delta t$ (av) s

| | | | | | | | | | | |
|---|---|---|---|---|---|---|---|---|---|---|
| 2.43 | 2.47 | 2.50 | 2.52 | 2.55 | 2.57 | 2.58 | 2.60 | 2.65 | 2.71 | 2.79 |
| 2.87 | 2.94 | 3.10 | 3.28 | 3.51 | 3.61 | 3.80 | 3.96 | 4.07 | 4.46 | |

$\Sigma = 64.0$ s

Power absorbed per interval:

$$\Delta P = (h_i)\Delta Q \rho g = 1414/\Delta t \qquad (4.94)$$

ΔP, kW

| 0.58 | 0.57 | 0.57 | 0.56 | 0.55 | 0.55 | 0.55 | 0.54 | 0.53 | 0.52 | 0.51 |
|------|------|------|------|------|------|------|------|------|------|------|
| 0.49 | 0.48 | 0.46 | 0.43 | 0.40 | 0.39 | 0.37 | 0.36 | 0.35 | 0.32 | |

Total = 10.08 kW

**Comment**

The time it takes for the axial pump to reach the operating point is much less than the time for the centrifugal pump. However, the power required is considerably higher. Depending on the system, this may not be acceptable—especially with the larger current spike at start-up.

**4.12.8** Comparison of the behavior of a centrifugal pump working at constant speed and variable speed with a pump operating at constant head and variable discharge. In the case of the constant-speed motor, throttling regulates the discharge. The pump characteristics are shown in Figure 4-88.

Data: Constant head = 15 m: Operating N = 875 rpm: Required discharge = 150 liters/s

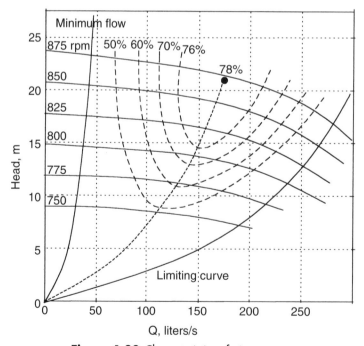

**Figure 4-88** Characteristics of given pump.

## Solution

1. Constant-speed motor

   From Figure 4-88 at a speed of 875 rpm, values corresponding to Q = 150 l/s are: H = 19.25 m and an efficiency = 77% (approx.)

   Power requirement of the pump = (150)(21.75)(9810)/(0.77)(1000) = 41,565 W = 41.6 kW

   Power absorbed at the throttling valve = (21.7 − 15)(150)(9810)/(1000) = 9933 W = 9.9 kW

2. Variable-speed motor

   Here the pump is slowed to deliver the exact Q required.

   Referring to Figure 4-88: speed = 813 rpm (approx.) and efficiency = 0.76 (approx.)

   Power requirement of the pump = (150)(15)(9810)/(1000)(0.762) = 29,043 W = 29.0 kW

## Comment

If the required conditions were closer to the design point, the power wasted at the valve would have been considerably reduced.

**4.12.9** Estimate the dimensions of an acceptable axial-flow pump with diffuser vanes to conform to the following data:

$$\text{Effective head} = 5.0\,\text{m, discharge} = 1500\,\text{l/s}$$

Also, determine the axial thrust.

## Solution

The empirical dimensionless groups for axial-flow pumps will be used for this problem. We will assume $\eta_0 = 0.85$ for this problem; this is a good average value for pumps of this type. As a starting point, a value of diameter ratio $(d_{HUB}/d_{TIP}) = 0.4$ is assumed. This is the lower end of the usual range 0.4 to 0.55. An initial value of either $d_{HUB}$ or $d_{TIP}$ must be assumed, which may be corrected to conform to the flow ratio range.

Assume $d_{HUB} = 0.5$ m initially.

Substituting values in the flow ratio group:

$$Q/[(\pi/4)(d_{TIP}^2 - d_{HUB}^2)(2gH)^{0.5}] \tag{4.95}$$

The value of the flow ratio = 0.15.

This is outside the suggested range of 0.25 to 0.6. A new value of $d_{HUB}$ is assumed; say $d_{HUB} = 0.3$ m. The new value of the flow ratio = 0.41. This is acceptable.

$$\text{The value of } d_{TIP} = 0.75\,\text{m.}$$

Using the value of $d_{TIP}$ in the speed ratio group $[(\pi d_{TIP}N/60)/(2gH)^{0.5}]$, which must lie between 2.0 and 2.7, we obtain N = 504 to 681 rpm. The inlet and outlet angles at hub and tip will be a function of N.

Average axial-flow velocity:

$$c_{AXIAL} = 1.5/[(\pi/4)(d_2^2 - d_1^2)] = 4.06\,\text{m/s} \qquad (4.96)$$

With regard to the velocity triangles of Figure 4-41, the absolute velocity vector at entry $c_1 = c_{AXIAL}$ because of zero prerotation at both the tip and hub of the blade. There is, of course, a slight induced prerotation, but this may be ignored.

The blade velocities are:

At tip $u_{TIP} = \pi d_{TIP}N/60 = 19.7$ or $26.6\,\text{m/s}$ for N = 504 or 681 rpm.

At hub $u_{HUB} = u_{TIP}(d_{HUB}/d_{TIP}) = 7.9$ or $10.6\,\text{m/s}$ for N = 504 or 681 rpm.

Calculation of angles:
Inlet angle at tip:

$$\beta_{1TIP} = \tan^{-1}(c_{AXIAL}/u_{TIP})$$

$$= 11.6° \text{ for 504 rpm and } 8.7° \text{ for 681 rpm} \qquad (4.97)$$

$$\text{Outlet angle at tip} = \beta_{2TIP} = \tan^{-1}c_{AXIAL}/(u_{TIP} - c_{U2TIP}) \qquad (4.98)$$

$c_{U2TIP}$ must be calculated from the definition of hydraulic efficiency:

$$c_{U2TIP} = (Hg)/[(\eta_{HYD})(u_{TIP})] = [(5)(9.81)]/[(0.85)(19.7)] = 2.93\,\text{m/s at 504 rpm} \qquad (4.99)$$

and

$$c_{U2TIP} = (Hg)/[(\eta_{HYD})(u_{TIP})] = [(5)(9.81)]/[(0.85)(26.6)] = 2.17\,\text{m/s at 681 rpm} \qquad (4.100)$$

Substituting in Equation (4.98):

$$\beta_{2TIP} = \tan^{-1}(4.06)/(19.7 - 2.93) = 13.6° \text{ at 504 rpm}$$

and

$$\beta_{2TIP} = \tan^{-1}(4.06)/(26.6 - 2.17) = 9.4° \text{ at 681 rpm}$$

Similarly:

$$\beta_{1HUB} = \tan^{-1}(c_{AXIAL}/u_{HUB}) = 27.2° \text{ for } 504 \text{ rpm and } 21.0° \text{ for } 681 \text{ rpm} \qquad (4.101)$$

$$\text{Outlet angle at hub} = \beta_{2HUB} = \tan^{-1}(c_{AXIAL}/u_{HUB} - c_{U2\ HUB}) \qquad (4.102)$$

$$c_{U2\ HUB} = (Hg)/[(\eta_{HYD})(u_{HUB})] = [(5)(9.81)]/[(0.85)(7.9)] = 7.30 \text{ m/s at } 504 \text{ rpm} \quad (4.103)$$

and

$$c_{U2\ HUB} = (Hg)/[(\eta_{HYD})(u_{HUB})] = [(5)(9.81)]/[(0.85)(10.6)] = 5.44 \text{m/s at } 681 \text{ rpm} \quad (4.104)$$

Substituting in Equation (4.87):

$$\beta_{2HUB} = \tan^{-1}(4.06)/(7.9 - 7.30) = 81.6° \text{ at } 504 \text{ rpm} \qquad (4.105)$$

$$\beta_{2HUB} = \tan^{-1}(4.06)/(10.6 - 5.44) = 38.2° \text{ at } 681 \text{ rpm} \qquad (4.106)$$

**4.12.10** Demonstrate the advantage of using a four-stage pump over a single centrifugal pump in terms of disk friction power loss for the following operating data—total head: 100 m, flow rate: 25 liters/s, speed: 1450 rpm.

Overall efficiency for both pumps = 0.83.

The fluid pumped is water at 20° C.

### Solution
The specific speed of the pump

$$N_S = (NQ^{1/2}/H^{3/4}) = [(1450)(0.025)^{1/2}]/(100^{3/4}) = 7.3 \qquad (4.107)$$

The maximum value of $N_S$ for what is regarded as the upper limit, in SI units, of the low specific speed range of centrifugal pumps is approximately 19. Therefore, this pump falls in the lower range. We first assume a value for the diameter of a radial impeller for the single-stage pump, as 25 cm (0.25m). To calculate the disk friction power loss, we will use Equation (4.13):

$$(P_{DF}/P_W) = 7800/N_S^{5/3}$$

We recall that the units of $N_S$ in Equation (4.13) are: rpm, US gpm, and ft.

$$\therefore \qquad N_S = 374.$$

Substituting in Equation (4.13):

$$(P_{DF}/P_W) = 0.40 \tag{4.108}$$

$$P_W = (9810)(100)(0.025) = 24{,}525\,W = 24.5\,kW \tag{4.109}$$

$$\therefore \qquad P_{DF} = (0.40)(24.5) = 9.81\,kW \tag{4.110}$$

The diameter of one stage of the four-stage pump is given by:

$$(N_1/N_4)^2 = (D_4/D_1)^2(H_1/H_4) \tag{4.111}$$

Substituting values:

$$(1450/1450)^2 = (D_4/0.25)^2(100/25) \text{ and } D_4 = 0.125\,m \tag{4.112}$$

Repeating the calculation for one stage:

$$N_S = 1057 \text{ and } (P_{DF}/P_W) = 0.071 \tag{4.113}$$

$$P_W = (9810)(25)(0.025) = 6130\,W = 6.13\,kW \tag{4.114}$$

$$\therefore\ P_{DF} = (0.071)(6.13) = 0.435\,kW/stage. \text{ The total } P_{DF} = 1.74\,kW \tag{4.115}$$

This may be compared with 9.81 kW for the single-stage pump.

## 4.13 References

Busemann, A., *Das Forderhohenverhaltniss Radialer Krieselpumpen mit Logarithmischilarigen Schaufeln.* ZAMM 8, 372–384 (1928).

Föttinger, H., *Eine neue lösing des Schiffsturbinenproblems.* Jahrb. D. Schiffbautechn. Ges. 11 (1910) s .157.

Karassik, I.J., Krutzsch,W.C., Fraser, W.H., and Messina, J.P., *Pump Handbook.* McGraw-Hill, New York (1976).

Pfleiderer, C., *Die Krieselpumpen für Flüssigkeiten und Gase.* Springer-Verlag (1961), 5 Auflage, 8, S., 422 pp.

Round, G.F., Valavaara, V.K., and Peng, L., *Rotary Pump.* U.S. Patent No. 5,658,138, August 19, 1997.

Stanitz, J.D., "Some theoretical aerodynamic investigations of impellers in radial and mixed flow centrifugal compressors." *Trans. ASME* 74, 473–497 (1952).

Stepanoff, A.J., *Centrifugal and Axial Flow Pumps. Theory, Design and Application*. John Wiley & Sons, New York, 462 pp. (1957).

Stodola, A., *Steam and Gas Turbines*. McGraw-Hill, New York (1927).

Varley, F.A., "Effects of impeller design and roughness on the performance of centrifugal pumps." *Proc. Inst. Mech. Eng.* 175, 955–989 (1961).

Wiesner, F.Y., "A review of slip factor for centrifugal impellers." *Trans. ASME* 89, 558–572 (1967).

# SOME ASPECTS OF DESIGN

## 5.1 General Remarks

By now it should be self-evident that the designs of turbines and pumps have much in common, especially as far as the runners and impellers are concerned. Pumps, generally speaking, may be regarded as turbines operating in reverse. Of course, this is not always true. There is no pump equivalent to a Pelton wheel, for example. Inlet and outlet elements and connections differ considerably, although pump volutes and turbine scroll cases have many similarities.

In the remarks that follow, it will be assumed that whatever design procedures apply to pump impellers, the same procedures may be applied to turbine runners. A number of empirical rules based on performance data have been acquired over many years; these will be dealt with in this chapter and Chapter 6.

## 5.2 Application to Flow

### 5.2.1 Axial-flow Design

Over a period of time, a number of empirical dimensionless groups have been developed that have been found to be useful in the design of axial-flow pumps and turbines. These groups should not be applied as hard and fast rules; rather, they are useful guidelines to follow. The range of suggested values are on the right-hand side of each equation. They are:

**Speed Ratio:**

$$U_2/(2gH)^{0.5} = 2.0 \text{ to } 2.7 \qquad (5.1)$$

$H =$ Delivered head for pump

$\quad =$ Effective head across turbine

**Flow Ratio:**

$$Q/[(\pi/4)(d_2^2 - d_1^2)(2gH)^{0.5}] = 0.25 \text{ to } 0.6 \tag{5.2}$$

$Q$ = Volumetric flow rate

$d_2$ = OD

$d_1$ = ID

**Diameter Ratio:**

$$(d_1/d_2) = 0.4 \text{ to } 0.55 \tag{5.3}$$

In addition, the number of blades should be in the range 3 to 5.

# 5.3 Axial and Radial Thrusts in Pumps and Turbines

## 5.3.1 Axial

First, considering pumps: for single-entry impellers, two components must be considered:

1. The change in momentum of the fluid turning through 90°

$$T_A = T_1 - T_2 \tag{5.4}$$

T designates thrust

2. A thrust due to a pressure difference axially across the impeller:

$$T_3 = (p_{atm} - p_0)A_{SH} \tag{5.5}$$

$p_{atm}$ = atmospheric pressure on the free end of the shaft

$p_0$ = absolute pressure at the eye inlet of the impeller

$A_{SH}$ = shaft area

## 5.3.2 Closed Single-entry Centrifugal Impellers

All impellers and runners are subjected to axial thrust. The resultant force tends to move the impeller or runner away from the suction side because of the difference in static pressure on both sides of the impeller/runner. In addition, if the impeller/runner is mounted vertically, the thrust due to the weight of the rotating mass must be added. Figure 5-1 shows the pressure distributions on an impeller of a single-suction impeller.

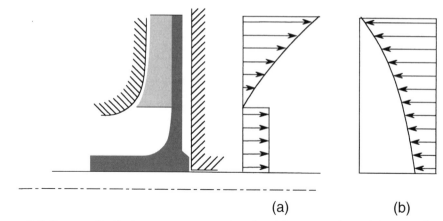

**Figure 5-1** Pressure distribution across an open impeller: (a) suction side; and (b) rear of impeller.

The pressure at the discharge end of the impeller is $H_2\rho g$. The static pressure is:

$$H_2 = H - c_{u2}^2 \tag{5.6}$$

The force generated by the pressure difference between front and rear of the impeller is given by:

$$(p_2 - p_0) = (\pi/4)(D_2 - D_0) \tag{5.7}$$

where:

2 refers to the rear

1 refers to the suction

The change of momentum, (mass flow rate) $\times c_0$, gives a force acting in the opposite direction. $c_0$ is the axial fluid velocity entering the impeller. The resultant axial thrust, acting toward the suction side, is:

$$(p_2 - p_0) = [(\pi/4)(D_2 - D_0)] - (\text{mass flow rate}) \times c_0 \tag{5.8}$$

The axial thrust can only be partially balanced by applying sealing rings on the hub and connecting the back shroud with the suction side. This may be accomplished with a series of holes connecting the suction side with the space at the rear of the impeller. Either the rear space is sealed so that the space is pressurized, or alternatively a series of radial vanes are fixed to the back of the impeller, in effect providing a small centrifugal pump at the rear. Figure 5-2 illustrates these arrangements.

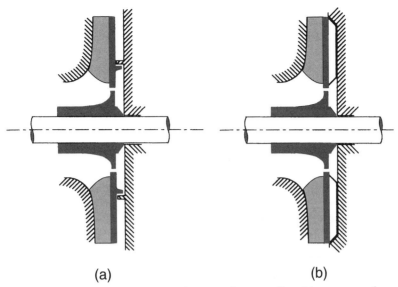

(a)                                          (b)

**Figure 5-2** (a) Sealed pressurized space at the rear of an impeller; (b) rear vanes for pressurizing space at rear.

### 5.3.3 Multistage Balancing of Single-entry Stages

Figure 5-3 shows a method of balancing a multistage pump with an even number of stages. The thrust from the stages to the left of the center bearing act from left to right and are exactly balanced by the thrust of the stages to the right of the center bearing acting from right to left.

### 5.3.4 Radial

The pressure distribution on the inlet circumference of the volute varies around the volute of a pump and the scroll case of a turbine due to the variation of area circumferentially. Radial thrust may be expressed as:

$$F_R = K_R \gamma D_2 b_2 (1 - Q/Q_0)^2 \qquad (5.9)$$

where:

$K_R$ = a factor that is a function of specific speed

$Q_0$ = flow rate at zero head

Agostinelli et al. (1960) have made measurements to determine $K_R$ as a function of specific speed.

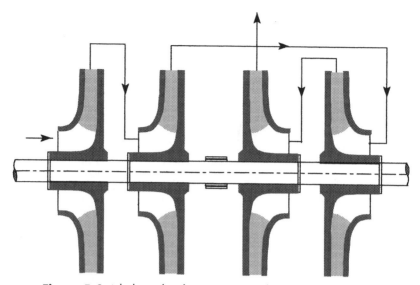

**Figure 5-3** A balanced multistage pump with single-entry stages.

## 5.4 Critical Speeds

For any rotating machine, as the speed of rotation is increased, it may be observed that at certain speeds the shaft of the machine will oscillate or vibrate. The vibration occurs because the machine is imbalanced as it rotates. Only rarely is a rotational machine perfectly balanced. Imperfections in manufacture of the shaft and "balanced" masses attached to the shaft mean that the geometric center of rotation and the center of mass do not coincide. Even a small mass imbalance may produce large centrifugal forces that are balanced by the spring action of the shaft, causing the system to vibrate excessively. The speeds at which this occurs are called critical speeds. If this condition is allowed to proceed, vibrations may cause large amplitudes to develop, leading to dangerously high stresses. Rubbing of parts may occur, and vibration may be transmitted through the machine to the foundations. It is therefore important that critical speeds be determined and that the running speed of the machine be at least 20% above or below a critical speed region. The critical speed occurs at the natural frequency of the shaft.

Critical speed analysis is founded on a principle formulated by Lord Rayleigh (1894)—the principle that the total energy of the system stays constant and by equating the maximum values of kinetic and potential energy, then the lowest natural frequency is obtained. Thus, if a shaft is loaded with a number of masses $m_1$, $m_2 \ldots m_n$, along its length and the corresponding maximum deflections are $y_1$, $y_2 \ldots y_n$, then the maximum potential energy is: $\frac{1}{2} m_1 g y_1 + \frac{1}{2} m_2 g y_2 + \ldots \frac{1}{2} m_n g y_n$. Similarly, the maximum kinetic energy is $\frac{1}{2} m_1 \omega^2 y_1^2 + \frac{1}{2} m_2 \omega^2 y_2^2 + \ldots \frac{1}{2} m_n \omega^2 y_n^2$.

Equating the energies and solving for $\omega$:

$$\omega = (g)^{1/2} \{\Sigma(my)/\Sigma(my^2)\} \quad \text{rads/s} \tag{5.10}$$

Alternatively,

$$N = (60/2\pi)(g)^{1/2} \{\Sigma(my)/\Sigma(my^2)\} \quad \text{rpm} \tag{5.11}$$

In this section, the concern will be mostly with pumps. Hydraulic turbines have low rotational speeds, and critical speeds are usually much higher. Two factors increase the critical speed of a turbine:

1. The mass of a turbine is much larger than the mass of the average pump. The runners create a large gyroscopic effect and resist any forces tending to change the direction of their axes. The effect tends to be pronounced near the bearings.
2. The bearings on high-capacity turbines are large, long, and rigid, causing a stiffening of the shaft and thus tending to resist deflections, raising the critical speed.

Two types of critical speed affect shafts: lateral critical speed and torsional critical speed. Lateral critical speeds are caused by mass imbalance on the shaft. Torsional critical speeds are a result of torque impulses that are transmitted to the shaft; these are caused by misalignments. The result of these torsional stresses, if they coincide with the natural frequency, or are even multiples of the natural frequency, can result in the buildup of high stresses with the possibility of shear fatigue failure.

Lateral critical speed and torsional critical speed will be considered separately in the discussion and examples that follow.

## 5.4.1 Lateral Critical Speed of an Unbalanced Simple Rotor

As an introductory example, we consider a disk of mass M mounted at the center of a shaft, K refers to the lateral stiffness of the shaft, and C is the viscous damping of the system. The equations that describe the unbalance of the rotor are:

$$M(d^2z/dt^2) + C(dz/dt) + Kz = M\omega^2 a \cos \omega t \tag{5.12}$$

$$M(d^2y/dt^2) + C(dy/dt) + Ky = M\omega^2 a \sin \omega t \tag{5.13}$$

where:

M = mass of disk
$\omega$ = angular rotational
a = distance between the center of the disk and its center of gravity
z, y = coordinates

The definition diagram is shown in Figure 5-4.

The radius, $r = O - O^1$, may be expressed as the complex quantity:

$$r = z + \mathbf{i}y \tag{5.14}$$

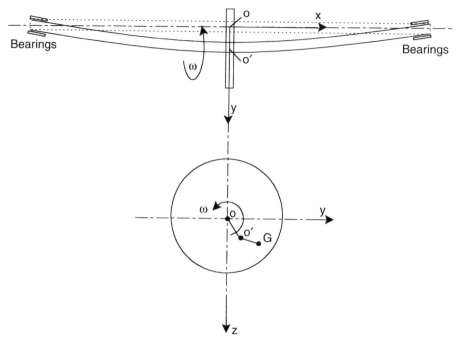

**Figure 5-4** Rotor with residual imbalance: (a) plan view; (b) side elevation.

This may be combined with Equations (5.12) and (5.13) to give:

$$M(d^2r/dt^2) + C(dr/dt) + Kr = M\omega^2 a\, e^{i\omega t} \tag{5.15}$$

The magnitude of the exciting force is the centrifugal force due to the eccentric disk. The disk rotates about the shaft axis, which is elastic and not about the bearing axis.

The steady-state solution of Equation (5.15) is:

$$r = Re^{i(\omega t - \phi)} \tag{5.16}$$

where:

$$R = M\omega^2 a/[(K - M\omega^2)^2 + C^2\omega^2]^{0.5} \tag{5.17}$$

$$\Re = R/a = \Omega^2/[(1 - \Omega^2)^2 + (2\zeta\,\Omega)^2]^{0.5} \tag{5.18}$$

where:

Frequency ratio, $\Omega = \omega/p$

$$p = (K/M)^{0.5}$$

$$\zeta = C/C_C$$

$$C_C = (KM)^{0.5}$$

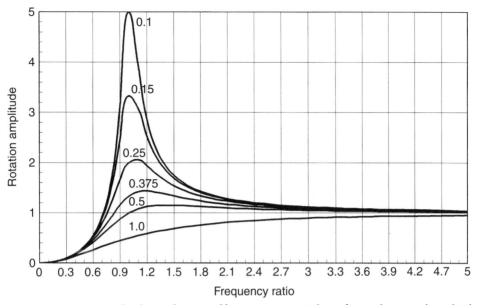

**Figure 5-5** Rotation amplitude as a function of frequency ratio. Values of $\zeta$ are shown on the individual curves.

$\phi$ is defined as:

$$\phi = \tan^{-1} 2\,\zeta\,\Omega/(1 - \Omega^2) \tag{5.19}$$

Using Equation (5.18), we see that Figure 5-5 is a plot of $\Re$ as a function of $\Omega$ for different values of $\zeta$.

Resonance occurs when $\omega = p$ and $\phi = 90°$ for an undamped rotor. For a rigid shaft in flexible bearings, there are two whirl modes: these are illustrated in Figure 5-6. Figure 5-6(a) shows a translatory whirl, and Figure 5-6(b) shows a conical whirl.

For a flexible shaft in flexible bearings there are two whirl modes (see Figure 5-7). Figure 5-7(a) shows a whirl about the horizontal axis, and Figure 5-7(b) shows a conical whirl with two bends in the shaft.

## 5.4.2 Multiple Disks

In Section 5.4.1, a simple rotor with one disk was considered. In practice, a shaft may have several disks for example, other disks, flywheels, and gears. Pumps that commonly have two bearings may have them combined with different impellers in a multiplicity of ways. Figure 5-8 illustrates some combinations.

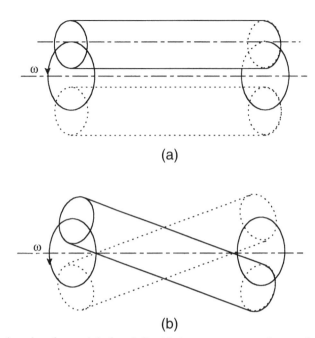

(a)

(b)

**Figure 5-6** Whirl modes of a rigid shaft with flexible bearings: (a) translatory whirl; (b) conical whirl.

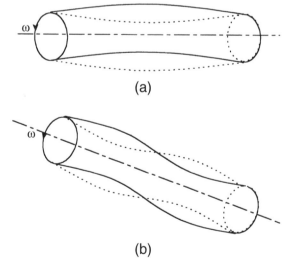

(a)

(b)

**Figure 5-7** Whirl modes of a flexible shaft with flexible bearings: (a) whirl about horizontal axis; and (b) conical whirl.

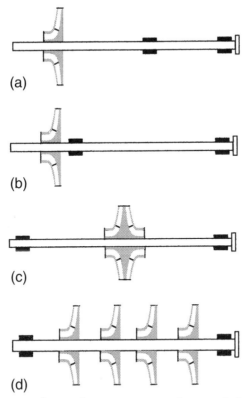

**Figure 5-8** Two-bearing, impeller combinations: (a) impeller outside both bearings; (b) impeller outside both bearings, with one bearing close to impeller; (c) double-suction impeller located at the center of the shaft, bearings symmetrically placed; and (d) multiple single impellers symmetrically positioned between two bearings.

One method of determining the fundamental frequency of multiple disk systems is to use the Rayleigh Equation (5.10). Another simple and convenient method was proposed by Dunkerley (1894). The method consists of assigning a mass to each part of the system such as an impeller, individual parts of the shaft, a flywheel, and so on. The critical speed of each component is then calculated. The critical speed of the entire system is determined by combining the critical speeds according to:

$$1/\omega^2 = 1/\omega_1^2 + 1/\omega_2^2 + 1/\omega_3^2 + \cdots 1/\omega_n^2 \qquad (5.20)$$

An immediate application difficulty associated with the application of Equation (5.11) is the determination of the deflections, y, of a shaft along its length when it is loaded in an arbitrary manner. This is usually done by a graphical method and is based on the area-moment method.

Unfortunately, this method is both lengthy and cumbersome. There are two other nongraphical ways of determining y:

1. By the use of singularity functions
2. By numerical integration

### 5.4.3 Use of Singularity Functions

A shaft is chosen which is simply supported, with a concentrated load F, not at the center. Figure A7-1 in the appendix illustrates the problem.

$$EI(d^4y/dx^4) = q = -F(x-a)^{-1} \quad 0 < x < L \tag{5.21}$$

The symbols have the meanings defined in the nomenclature. Integrating Equation (5.21) with respect to x, we obtain:

$$EI(d^3y/dx^3) = V = -F(x-a)^0 + C_1 \tag{5.22}$$

Integrating again:

$$EI(d^2y/dx^2) = M = -F(x-a)^1 + C_1x + C_2 \tag{5.23}$$

The boundary conditions are: at $x = 0$, $M = 0$, giving $C_2 = 0$. At $x = 1$, $M = 0$, and $C_1 = F(b/L)$. Substituting these in Equation (5.23) gives:

$$EI(d^2y/dx^2) = M = Fbx/L - F(x-a)^1 \tag{5.24}$$

Integrating Equation (5.24) twice:

$$EIy = M = Fbx^3/6L - F(x-a)^3/6 + C_3x + C_4 \tag{5.25}$$

$C_3$ and $C_4$ are evaluated from $y = 0$ at $x = 0$ and $y = 0$ at $x = L$. Thus

$$C_4 = 0 \text{ and } C_3 = (-Fb/6L)(L^2 - b^2) \tag{5.26}$$

The deflection reduces to:

$$y = (F/6EIL)[bx(x^2 + b^2 - L^2) - L(x-a)^3] \tag{5.27}$$

A number of standard beam solutions have been tabulated; see, for example, Shigley and Mischke (1989). Very often these can be combined to form more complex solutions. For example, the solution for y for a uniformly loaded shaft and the solution for y for a simple-supported, single-load shaft may be combined to obtain the y of a uniformly loaded shaft, with an additional single load placed at an arbitrary x on the shaft.

### 5.4.4 Solution by Numerical Integration

Mischke (1978) has also outlined an exact numerical method for application to the bending of stepped shafts. The method easily lends itself to computer programming. Referring to Figure A7-1, the method is as follows: a function is defined by the integral,

$$\phi = \int_0^x (M/EI)\,dx \tag{5.28}$$

where:

M = moment
E = modulus of elasticity
I = second moment of area

The slope (dy/dx) is given by:

$$\theta = (dy/dx) = \int_0^x (M/EI)\,dx + C_1 \tag{5.29}$$

where:

$C_1$ = slope at x = 0

The slope is a piecewise quadratic function. Simpson's rule is used to obtain the integral. The integral of Equation (5.28) is designated by:

$$\psi = \int_0^x \phi\,dx \tag{5.30}$$

The deflection y becomes:

$$y = \psi + C_1 + C_2 \tag{5.31}$$

$\theta$ and y may be written as:

$$\theta = K(\phi + C_1) \tag{5.32}$$

and

$$y = K(\psi + C_1 + C_2) \tag{5.33}$$

K depends on the units used.

Equations (5.32) and (5.33) become predictor equations. If the supports at $x = a$ and $x = b$ have zero deflection specified, then:

$$C_1 = (\psi_b - \psi_a)/(a - b) \tag{5.34}$$

$$C_2 = (b\psi_a - a\psi_b)/(a - b) \tag{5.35}$$

A forward-marching process may now be used to evaluate $\phi$ and $\psi$ by means of applying the trapezoidal rule to $\phi$ and Simpson's rule to $\psi$ as follows:

$$\phi_{i+2} = \phi_i + \frac{1}{2}[(M/EI)_{i+1} + (M/EI)_{i+1}](x_{i+2} - x_i) \tag{5.36}$$

$$\psi_{i+4} = \psi_i + \frac{1}{6}(\psi_{i+4} + 4\psi_{i+2} + \psi_i)(x_{i+4} - x_i) \tag{5.37}$$

Successive values of $\phi$ are computed using Equation (5.36) beginning at $x_1$ and ending at $x_n$; n is the number of (M/EI) values. In a similar way, values of $\psi$ are computed. After these numerical integrations, the constants $C_1$ and $C_2$ may be found. Equations (5.32) and (5.33) may then be solved for the slope and defection.

## 5.4.5 Torsional Critical Speed

A common vibration problem in turbomachines is a two-mass system in torsional vibration, shown schematically in Figure 5-9. The natural frequency of such a system is given by:

$$\omega = (60/2\pi)\{I_P E_S(J_1 + J_2)/(J_1 J_2 L)\}^{0.5} \tag{5.38}$$

where:

$\omega$ = cycles per minute
$I_P$ = moment of inertia of the shaft, $\pi d^4/32$
$d$ = shaft diameter
$E_S$ = shear modulus of elasticity
$J_1, J_2$ = mass moments of inertia
$L$ = shaft length

For stepped shafts it is convenient to choose a reference diameter and to convert parts of the shaft that are not of this diameter to equivalent lengths of the reference diameter. The equation used to

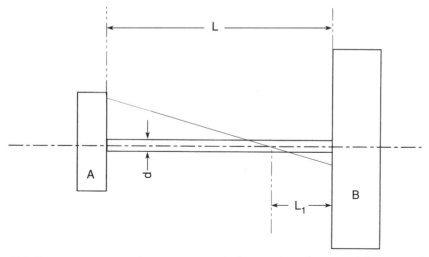

**Figure 5-9** Two-mass system undergoing torsional vibration: $L_1$ indicates nodal position; A = mass Moment of Inertia, $J_1$; B = mass Moment of Inertia, $J_2$.

do this is:

$$L_e = (D_e/D)^4 L \qquad (5.39)$$

An example of the use of Equation (5.39) is given at the end of this chapter.

## 5.5 Seals

Seals on rotating shafts fall into two categories:

1. Stuffing box seals
2. Mechanical seals

The aim of both these seals is to control leakage from a high-pressure liquid source through a gap, the space between the shaft and the casing, to the atmosphere. The following remarks apply principally to pumps, but the comments for mechanical seals apply equally well to turbines. The higher the pressure differential across the space dividing the liquid from atmosphere, the greater is the driving force and consequently the greater is the leakage. To help alleviate the problem, designers try to lower the differential pressure. For pumps, this is done using vanes on the rear of the impeller of a radial of mixed-flow centrifugal pump; these vanes generate a back-pressure. Alternatively, balancing holes are drilled through the impeller to increase the pressure on the rear surface. Each of these devices decreases the overall efficiency of the pump. Figures 5-2(a) and (b) have been used already, and they illustrate the point.

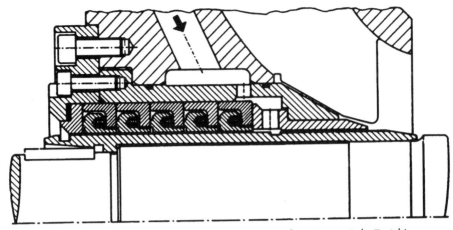

**Figure 5-10** A typical stuffing box. (Courtesy Sulzer Pumps, Ltd., Zurich)

Even when using both of these remedies, leakage still occurs and sealing is necessary. This may be done using both types of seal. The advantage of packed seals, stuffing boxes, is low initial cost. Two advantages of mechanical seals are lower operating costs and maintenance.

In a typical stuffing box as shown in Figure 5-10, there are five packing rings. This type of stuffing box could be used where there is very little leakage because of a small pressure differential. Figure 5-11 shows a pressure-balanced stuffing box with a barrier water ring for injection of water; this type would be used for a condensate pump. Other pressure-balanced stuffing boxes would use a lubricating or flushing liquid and a gland to hold the packing and maintain the desired compression pressure. The packing is intended to control leakage but not to eliminate it entirely. In fact, it is desirable to keep the packing lubricated, cooled, and sealed; otherwise the packing will burn. The amount of controlled leakage that will accomplish this is small; the leakage needs only be one drop every 1 to 2 seconds. The lubrication method depends on the pumped liquid; for example, if the liquid is clean and nonabrasive, the pumped liquid will lubricate the packing. When this is not the case, sealing liquid pressure must be kept above stuffing box pressure. One disadvantage of packed seals is that they tend to scar and mark the shaft.

The main difference between the use of packed seals and mechanical seals is determined by the amount of leakage past the seal.

Mechanical seals have other advantages over packed seals:

1. They can operate at much higher suction pressure.
2. Because no sealing fluid is needed, there is no product contamination.
3. Handling of corrosive liquids is easier because a buffer region may be integrated into the seal to prevent any fluid migration at all.

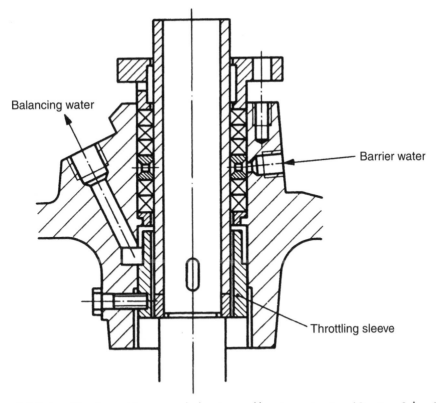

Balancing water

Barrier water

Throttling sleeve

**Figure 5-11** A stuffing box with pressure balancing and barrier water ring. (Courtesy Sulzer Pumps, Ltd., Zurich)

Figures 5-12 and 5-13 show two types of mechanical seal, Figure 5-12 is a single-acting seal and Figure 5-13 is a double-acting seal.

## 5.6 Cooling Seals

Pumps have a wide range of shaft speeds. A shaft speed of several hundred to several thousand rpm implies that a large amount of heat may be generated at the seal. Both cooling and flushing capability are important—cooling from the point of view of effective heat dissipation and flushing because of the potential of settled particulate matter. Without this capability, the seals may fail. Cooling and flushing of seals may be carried out in two ways:

1. *Internal circulation of pump fluid.* In this case, some of the discharge fluid is bypassed to the seal faces and then returned to discharge.
2. *External circulation of coolant.* In this case, if the pumped fluid is unsuitable as a coolant or when it needs to be filtered, an external piping system should be provided.

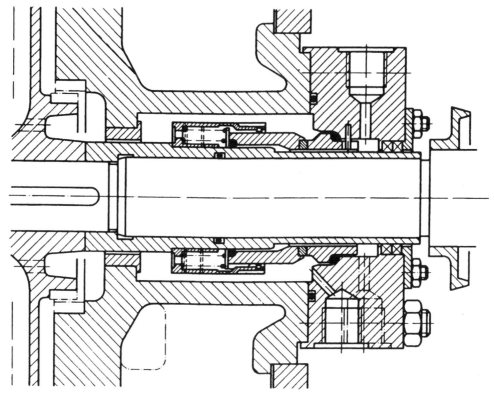

**Figure 5-12** A single-acting mechanical seal. (Courtesy Sulzer Pumps, Ltd., Zurich)

## 5.7 Glands

The function of a gland is to close off the end of a stuffing box or mechanical seal. Circulation or flush glands are designed to allow circulation of fluid for cleansing, cooling, and lubrication. Sometimes these glands incorporate both a drain and a vent.

## 5.8 Solved Problems

**5.8.1** Example of critical speed determination: critical speed of two impellers located on a two-bearing shaft.

Two impellers are located as shown on a shaft driven by a motor. The supports are self-aligning bearings. The system and the geometric data for the problem, together with the forces acting, are shown in Figure 5-14.

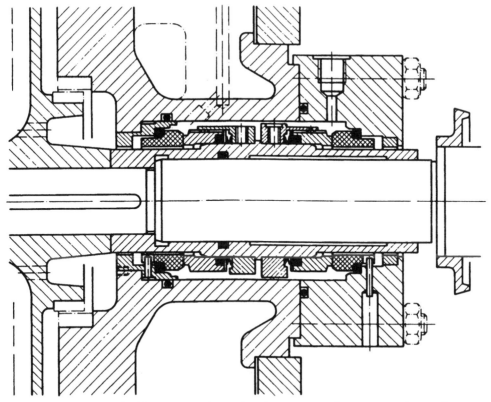

**Figure 5-13** A double-acting mechanical seal. (Courtesy Sulzer Pumps, Ltd., Zurich)

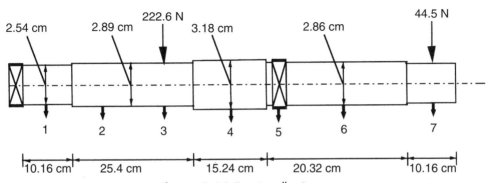

**Figure 5-14** Two-Impeller System.

Given:

$$E_S \text{ (carbon steel)} = 79.3 \text{ GPa}$$
$$\gamma \text{ (carbon steel)} = 76.5 \text{ kN/m}^3$$
$$d \text{ (shaft)} = 0.0508 \text{ m}$$
$$\text{Weight of impellers} = 222.6 \text{ N and } 44.5 \text{ N}$$

What is the lateral critical speed of the system?

## Solution

The method used to determine the required deflections is the superposition principle. The system is considered to be composed of a series of joined shafts of differing diameters and two loads due to the impellers, each of which causes deflection. These deflections are calculated separately and then added. The shaft sections are considered to be uniform beams, and the weight of a section of shaft is considered to act at the geometric center of the section. For convenience, the system is divided into seven sections. Measuring x—the distance along the shaft from the end of the left-hand side bearing—the deflection y at any x for a loaded shaft as in Appendix A-7 is given by:

$$y_{AB} = [(Fbx)/6E_SI](x^2 + b^2 - L^2) \tag{5.40}$$

$$y_{BC} = [Fa(L - x)/6E_SI](x^2 + a^2 - 2Lx) \tag{5.41}$$

For the first set of calculations, Equation (5.40) is applied at x(1) and x(2), and Equation (5.41) is applied to point x(3).

$$y = [(Wx)/24 E_SI] (2Lx^2 - x^3 - L^3) \tag{5.42}$$

$I = $ second moment of area about the axis of rotation
$E_S = $ modulus of elasticity
$L = $ length of shaft

Weight of shaft:

$$W = (\pi/4)d^2L\rho = (\pi/4)(0.1016)^2(0.4573)(76500) = 22.47 \text{ N}$$

$$I = (\pi/64)d^4 = (\pi/64)(0.1016)^4 = 5.255 \times 10^{-7}\text{m}^2$$

$$L = 0.4573 \text{ m}$$

y for the uniform shaft is calculated for three positions, corresponding to the centers of AB and BC and at point B. Substituting in Equation (5.42) at x(1) = 0.0508 m, x(2) = 0.1779 m, and x(3) = 0.3557 m, we obtain:

$$y[x(1)] = -4.088 \times 10^{-8} \text{ m: } y[x(2)] = -11.08 \times 10^{-8} \text{ m: } y[x(3)] = -7.636 \times 10^{-8} \text{ m}$$

**Table 5-1** Calculation of Critical Speed

|        | W(N)   | y (m)     | $y^2$        | Wy        | $Wy^2$     |
|--------|--------|-----------|--------------|-----------|------------|
| x(1)   | 4.01   | 3.81e−05  | 1.4516e−09   | 0.0001528 | 5.821e−09  |
| x(2)   | 7.57   | 1.19e−04  | 0.00000001   | 0.0009039 | 0.0000001  |
| x(3)   | 227.49 | 1.50e−04  | 0.00000002   | 0.034078  | 0.0000051  |
| x(4)   | 9.35   | 0.0000991 | 9.8208e−09   | 0.0009266 | 9.182e−08  |
| x(5)   | 4.9    | 0         | 0            | 0         | 0          |
| x(6)   | 4.9    | 0.0000965 | 9.3122e−09   | 0.0004729 | 4.563e−08  |
| x(7)   | 47.64  | 0.000198  | 0.00000004   | 0.0094327 | 0.0000019  |
| Σ      |        |           |              | 0.0459668 | 0.0000072  |

Critical speed, $N_C = (29.9)(\Sigma\, Wy/\Sigma\, Wy^2)^{\frac{1}{2}} = 2250$ rpm

The deflection at the center of the shaft at $x = 0.2287$ m is $y = -11.77 \times 10^{-8}$ m. The deflections at x(1), x(2), and x(3) for a shaft with a load located at x(2) are given by two equations:

$$y_{AB} = [(Fbx)/6EI](x^2 + b^2 - L^2) \tag{5.43}$$

$$y_{BC} = [Fa(L - x)/6EI](x^2 + a^2 - 2Lx) \tag{5.44}$$

Table 5-1 shows the results of the computations.

**5.8.2** For the vertical sewage pumping system shown in Figure 5-15, which consists of a stepped shaft, connecting a motor with its coupling to a submersible pump.

Given:

$J_1$ = mass moment of inertia of the motor coupling = 273.4 m-N-s$^2$

$J_2$ = mass moment of inertia of the pump = 76.1 m-N-s$^2$

$E_S$ = shear modulus of elasticity of carbon steel = 79.3 GPa

What are the fundamental torsional critical speed and location of the node or zero amplitude point?

**Solution**

First, the stepped shaft must be translated into a shaft of equivalent constant diameter. The reference diameter chosen is 2.5 cm. Equation (5.39) is used to calculate individual values of equivalent lengths of each section. The calculated values are shown in Table 5-2.

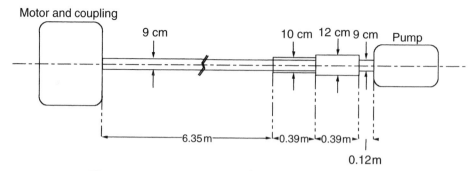

**Figure 5-15** Pump–motor combination on a stepped shaft.

**Table 5-2** Calculation of Equivalent Length

| L cm | D cm | $(d_e/d)^4$ | $L_e = L\,(d_e/d)^4$ |
|------|------|-------------|----------------------|
| 12   | 9    | 0.00595     | 0.0714               |
| 39   | 12   | 0.00188     | 0.0733               |
| 39   | 10   | 0.00391     | 0.0152               |
| 635  | 9    | 0.00595     | 3.7783               |

The total equivalent length is: $(0.0714 + 0.0733 + 0.0152 + 3.7783) = 3.938$ cm of 2.5 cm shaft.

The polar moment of inertia of the shaft, $I_P = (\pi/32)(2.5)^4 = 3.84$ cm$^4$ = $3.84 \times 10^{-8}$ m$^4$. Substituting values in Equation (5.38):

$$\omega = (60/2\pi)\{I_P E_S (J_1 + J_2)/(J_1 J_2 L)\}^{0.5}$$

$$= (60/2\pi)\{[(3.84 \times 10^{-8})(79.3 \times 10^9)(273.4 + 76.1)]/[(273.4 \times 76.1)(0.03938)]\}^{0.5}$$

$$= 344 \text{ cycles/min.}$$

# 5.9 References

Agostinelli, A., Nobles, D., and Mockridge, C.R., "An experimental investigation of radial thrust in centrifugal pumps." *Trans. ASME, J. Eng. Power* Ser. A, 82, 120–126 (1960).

Dunkerley, S., On the whirling of vibration of shafts. *Phil. Trans. Royal Soc.*, Series A, 185 (1894).

Karassik, I.J., Krutzsch, W.C., Fraser, W.H., and Messina, J.P., *Pump Handbook.* McGraw-Hill, New York (1976).

Mischke, C.R., An exact numerical method for determining the bending deflection and slope of stepped shafts. *Proc. Annual Winter Meeting ASME*, San Francisco (December 1978).

Shigley, J.E., and Mischke, C.R., *Mechanical Engineering Design* (5th Ed.). McGraw-Hill, New York (1989).

Strutt, J.W. *Baron Rayleigh, Theory of Sound* (2nd Ed. rev.). Dover Publications, New York (1945).

# DESIGN OF IMPELLERS AND RUNNERS OF SINGLE AND DOUBLE CURVATURE

## 6.1 General Remarks on Design of Runners and Impellers

The shape of a flow passage for either a runner or an impeller depends on the H, Q, and N—represented by the specific speed. These in turn are functions of:

1. The outlet tip speed, $u_2$, and the outlet meridional velocity, $c_{m2}$
2. The outlet blade angle, $\beta_2$
3. The number of blades, z
4. The ratio, $c_{u2}/c_{u3}$
5. The diameter ratio, $d_1/d_2$

## 6.2 Single-Curvature Design

As an example of the design problems and techniques used for single-curvature blades, the problem of centrifugal pump design will be used. The methods will be equally applicable to turbine runners.

For a centrifugal pump, head may be maintained at the same value with a smaller $u_2$ and a smaller $d_2$ with the same rotational speed N by increasing $\beta_2$ and z. Thus, the problems of design and associated calculations may result in several solutions, but they will not be of equal value in terms of efficiency and possibly production costs.

### 6.2.1 Meridional Velocities, Inlet Diameter, and Inlet Angle

Meridional velocities are calculated from:

$$c_{m1} = K_{cm1} \times (2gH)^{0.5} \tag{6.1}$$

$$c_{m2} = K_{cm2} \times (2gH)^{0.5} \tag{6.2}$$

$K_{cm1}$; $K_{cm2}$ are velocity coefficients.

$c_{m1}$; $c_{m2}$ are meridional velocities

Values of $K_{cm1}$ and $K_{cm2}$ given in a plot by Stepanoff (1957) have been modified for units and plotted on an arithmetic basis rather than a logarithmic basis by Stepanoff. These values will be used in the example given in Section 6.3.

An empirical Russian equation that is sometimes used for the initial calculation of an inlet diameter is:

$$d0 = (4.0 - 4.5)(Q/N)^{1/3} \qquad (6.3)$$

$\beta_1$, the inlet angle, is calculated from:

$$\tan \beta_1 = c_{m1}/u_1 \qquad (6.4)$$

where:

$$u_1 = (N/60)\pi d_1 \qquad (6.5)$$

Experiments have shown that Equation (6.4) does not give the volumetric flow rate at the best efficiency point. $\beta_1$ must be increased by the value of the incidence angle $\delta_1$. This is $= 2\text{--}6°$. Thus,

$$\beta_1^l = \beta_1 + \delta_1 \qquad (6.6)$$

## 6.2.2 Tip Impeller Velocity, $u_2$, and Outlet Diameter, $d_2$

The theoretical head for a centrifugal pump with an infinite number of blades was given in Chapter 4 as:

$$H_{th}(\infty) = (1/g)(u_2 c_2 \cos \alpha_2 - u_1 c_1 \cos \alpha_1) \qquad (6.7)$$

From the outlet velocity triangle, it follows that:

$$c_{u2} = u_2 - c_{m2}/\tan \beta_2 \qquad (6.8)$$

Substitution of Equation (6.8) in Equation (6.7) yields:

$$gH_{th}(\infty) = u_2(u_2 - c_{m2}/\tan \beta_2) - u_1 c_{u1} \qquad (6.9)$$

or

$$u_2^2 - u_2 (c_{m2}/\tan \beta_2) = gH_{th}(4) + u_1 c_{u1} \qquad (6.10)$$

Hence:

$$u_2 = c_{m2}/2 \tan \beta_2 + [(c_{m2}/2 \tan \beta_2)^2 + gH_{th}(4) + u_1 c_{u1}]^{0.5} \qquad (6.11)$$

Usually, $c_{u1} = 0$ because of axial entry of the fluid. Equation (6.11) becomes:

$$u_2 = c_{m2}/2 \tan \beta_2 + \{(c_{m2}/2 \tan \beta_2)^2 + (gH/\eta_h)(1 + Cp)\}^{0.5} \qquad (6.12)$$

where:

(1 + Cp) is evaluated from the Pfleiderer correction for a finite number of blades.

$$(1 + Cp) = 2(\psi/z)[1/(1 - (d_1/d_2))^2] \qquad (6.13)$$

where:

$$\psi = k(1 + \sin \beta_2)(d_1/d_2) \qquad (6.14)$$

The value of $k = 1$ or 1.2, depending on whether or not the pump has guide vanes. For pump without guide vanes $k = 1.2$.

The outlet diameter is given by:

$$d_2 = [(60)(u_2)]/[(\pi)(N)] \qquad (6.15)$$

## 6.2.3 Inlet Areas and Impeller Widths

Inlet areas are given by:

$$A_1 = \psi Q^1/c_{m1} \qquad (6.16)$$

$\psi$ = a coefficient of constriction. This allows for reduction of flow blade area because of the presence of the blades.

$$Q^1 = Q/0.96 \qquad (6.17)$$

The value 0.96 is a commonly used value for volumetric efficiency.

Impeller widths at inlet and outlet are calculated from:

$$b_1 = A_1/\pi d_1 \qquad (6.18)$$

and

$$b_2 = A_2/\pi d_2 \qquad (6.19)$$

### 6.2.4 Dimension Calculations, Continuity Adjustments

There are three principal methods for designing blades:

1. Circular arc method
2. Point-by-point method
3. Conformal representation

The first of these methods may be carried out by a single-arc or double-arc method. Both methods are less accurate than the second—the point-by-point method. The point-by-point method, given as a design example in Section 6.3, was originally introduced by C. Pfleiderer (1957) and is illustrated in Figure 6-1.

Referring to the figure, the angle increment $d\theta$ is given by:

$$d\theta = dr/(r \tan \beta) \qquad (6.20)$$

Integrating between $r_2$ and r gives $\theta$, expressed in degrees:

$$\theta = (180/\pi) \int_{r2}^{r} (r \tan \beta)^{-1} dr \qquad (6.21)$$

When the initial blade design is done, a final check on the dimensions of the passages between the blades must be made. In effect, this is a check for flow continuity. The normals to the blade surfaces are drawn by means of inscribed circles, as illustrated by Figures 6-2 and 6-3. A given cross section has the shape of a trapezium of height $\Delta e$ and breadth b. A similar figure to Figure 6-2 appeared in Chapter 4. Figure 6-3 illustrates the type of deviation from the trapezoidal shape. The blade shape is then corrected for each cross section.

## 6.3 Example of Design—Blade of Single Curvature

The following is a worked example, applying the above principles and equations, to the design of the impeller of a centrifugal pump with blades of single curvature.

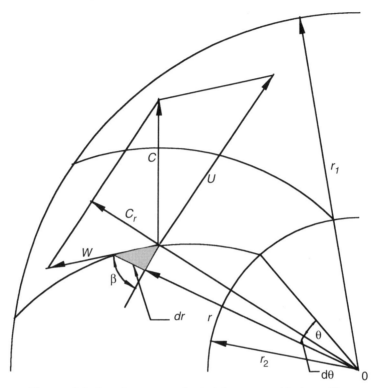

**Figure 6-1** Point-by-point method of determining blade profile.

Data: flow rate, $Q = 0.2$ m$^3$/s, required head, $H = 25$ m, volumetric efficiency, $\eta_v = 0.96$, overall efficiency, $\eta_O = 0.78$. Fluid: water, the pump to be driven by a three-phase 60 Hz supply. A range of impeller speeds is assumed initially 1000–2000 rpm (see Table 6-1).

## Solution

Other initial assumptions:

1. We assume that the impeller has eight blades. This can be changed later if desired.
2. An impeller ID will be assumed that lies between $d_0$ and $d_{HUB}$; once these have been calculated, the value of $d_1$ may be determined.
3. A diameter ratio of 0.5 ($d_1/d_2$) is assumed initially.

Power into the pump shaft:

$$P_{shaft} = \gamma HQ/\eta_O = (9806)(25)(0.2)/(0.78) = 62.86 \, \text{kW} \qquad (6.22)$$

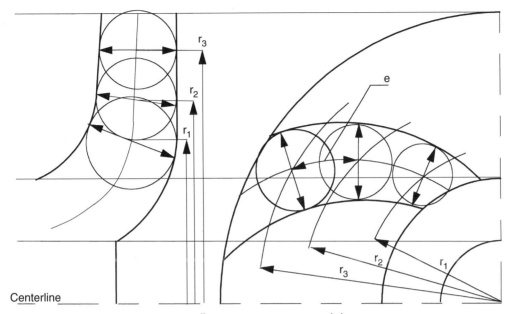

**Figure 6-2** Impeller passage cross-sectional determination.

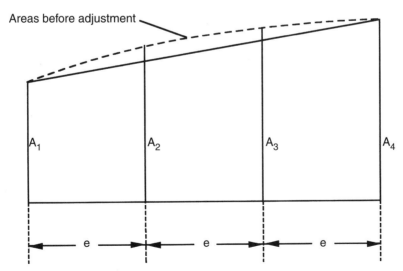

**Figure 6-3** Variation of impeller passage cross-sectional area with length.

**Table 6-1** Equations for data:
$K_{cm1} = 0.001923(N_S) + 0.0615$
$K_{cm2} = 0.001923(N_S) + 0.0615$

| N rpm | $N_S$ | $K_{cm1}$ | $K_{cm2}$ | $c_{m1}$ m/s | $c_{m2}$ m/s |
|-------|-------|-----------|-----------|--------------|--------------|
| 1000 | 40 | 0.1384 | 0.1670 | 3.065 | 3.699 |
| 1100 | 44 | 0.1461 | 0.1743 | 3.235 | 3.859 |
| 1200 | 48 | 0.1538 | 0.1815 | 3.406 | 4.018 |
| 1300 | 52 | 0.1615 | 0.1887 | 3.576 | 4.178 |
| 1400 | 56 | 0.1692 | 0.1959 | 3.746 | 4.338 |
| 1500 | 60 | 0.1769 | 0.2031 | 3.917 | 4.498 |
| 1600 | 64 | 0.1846 | 0.2103 | 4.087 | 4.657 |
| 1700 | 68 | 0.1923 | 0.2176 | 4.257 | 4.817 |
| 1800 | 72 | 0.2000 | 0.2248 | 4.428 | 4.977 |
| 1900 | 76 | 0.2076 | 0.2320 | 4.598 | 5.137 |
| 2000 | 80 | 0.2153 | 0.2392 | 4.768 | 5.297 |

Specific speed is calculated from:

$$N_S = N(Q)^{0.5}/(H)^{0.75} \tag{6.23}$$

where:

N = rpm

The values are shown in Table 6-1.

A plot of $K_{cm1}$ and $K_{cm2}$ versus specific speeds Ns given by Stepanoff (1957) is shown in modified form in Figure 6-4.

Inlet and outlet velocities of the impeller are calculated from:

$$c_{m1} = K_{cm1} \times (2gH)^{0.5} \tag{6.24}$$

$$c_{m2} = K_{cm2} \times (2gH)^{0.5} \tag{6.25}$$

Values $c_{m1}$ and $c_{m2}$ are given in Table 6-1.

Torsional strength of high-grade carbon steel, $\tau = 50\text{–}60\,\text{kp/mm}^2$

Taking the lower value: $\tau = 500\,\text{kp/cm}^2$ and allowing for a shaft keyway, take $\tau = 450\,\text{kp/cm}^2$. The shaft diameter is calculated from:

$$d_{shaft} = [(360{,}000)(P_{shaft})/(450)(N)]^{1/3} \tag{6.26}$$

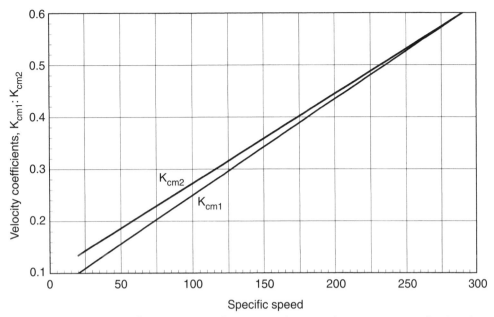

**Figure 6-4** Velocity coefficients $K_{cm1}$ and $K_{cm2}$ as a function of $N_S$. From: Centrifugal and Axial Flow Pumps by A.J. Stepanoff. Copyright ©1957 by John Wiley & Sons, Inc. This material is used by permission of John Wiley & Sons, Inc.

where:

$P_{shaft}$ = metric hp (1 metric hp = 746 W)
$N$ = rpm

For this example, substituting in Equation (6.26):

$$d_{shaft} = [(360,000)(84.26)/(450)(N)]^{1/3} = 40.7(N)^{1/3} \tag{6.27}$$

Values of $d_{shaft}$ were calculated for the range of N considered and are given in Table 6-2. These values calculated from Equation (6.26) are in cm. The hub diameter on the inlet side is usually taken to be:

$$d_{hub} = 1.4 \times d_{shaft} \tag{6.28}$$

A common volumetric efficiency for centrifugal pumps is 96%. Therefore, the design Q becomes:

$$Q^1 = Q/0.96 \tag{6.29}$$

The inlet cross-sectional area is:

$$A_0 = Q^1/c_{m1} \qquad (6.30)$$

It is usual to increase this by 5%.

The total inlet cross-sectional area is:

$$A_0^1 = A_0 + A_{HUB} \qquad (6.31)$$

Values of $A_0^1$ are given in Table 6-2. The diameter of the impeller eye inlet is:

$$d_0 = (4A_0^1/\pi)^{0.5} \qquad (6.32)$$

If there is forward extension of the impeller, this will be reduced.

Blade velocity at inlet is:

$$u_1 = (\pi d_1 N)/60 \qquad (6.33)$$

Water enters the impeller freely, that is, $\alpha_0 = 90°$, so:

$$\tan \beta_1 = c_{m1}/u_1 \qquad (6.34)$$

The flow angle of incidence at inlet is $\delta_1$. Thus, the flow angle becomes:

$$\beta_1' = \beta_1 + \delta_1 \qquad (6.35)$$

The values of $A_0^1$, $d_0$, $d_{HUB}$, $c_{m2}$ and $\beta_1^1$ are given in Table 6-2.

Equation (6.12) is now used to calculate values of the outlet angle $\beta_2$. That is,

$$u_2 = c_{m2}/2\tan \beta_2 + \{(c_{m2}/2\tan \beta_2)^2 + (gH/\eta_h)(1 + Cp)\}^{0.5} \qquad (6.36)$$

In this equation, Cp is evaluated by means of Equations (6.13) and (6.14). Values of $c_{m2}$ are presented in Table 6-1.

The hydraulic efficiency is defined as:

$$\eta_h = (\eta_O)/[\eta_v \times \eta_m] \qquad (6.37)$$

where:

$\eta_O$ = overall efficiency

**Table 6-2**

| N rpm | $d_{SHAFT}$ cm/s | $d_{HUB}$ cm/s | $Q^1$ $m^3/s$ | $A_0^1$ $m^2$ | $d_0$ cm | $u_1$ m/s | $\beta_1^1$ | $u_2$ cm/s |
|-------|------------------|----------------|---------------|---------------|----------|-----------|-------------|------------|
| 1000 | 4.48 | 6.27 | 0.208 | 0.079 | 31.8 | 7.85 | 24.3 | 15.71 |
| 1100 | 4.34 | 6.08 | 0.208 | 0.075 | 30.9 | 8.64 | 23.5 | 17.28 |
| 1200 | 4.21 | 5.89 | 0.208 | 0.071 | 30.2 | 9.42 | 22.9 | 18.85 |
| 1300 | 4.1 | 5.74 | 0.208 | 0.068 | 29.4 | 10.21 | 22.3 | 20.42 |
| 1400 | 4 | 5.60 | 0.208 | 0.065 | 28.7 | 11.00 | 21.8 | 22.00 |
| 1500 | 3.91 | 5.47 | 0.208 | 0.062 | 28.1 | 11.78 | 21.4 | 23.56 |
| 1600 | 3.83 | 5.36 | 0.208 | 0.060 | 27.5 | 12.57 | 21 | 25.13 |
| 1700 | 3.75 | 5.25 | 0.208 | 0.057 | 27.0 | 13.35 | 20.7 | 26.7 |
| 1800 | 3.68 | 5.15 | 0.208 | 0.055 | 26.4 | 14.14 | 20.4 | 28.27 |
| 1900 | 3.61 | 5.05 | 0.208 | 0.053 | 25.9 | 14.92 | 20.1 | 29.85 |
| 2000 | 3.55 | 4.97 | 0.208 | 0.051 | 25.5 | 15.71 | 19.9 | 31.42 |

**Table 6-3**

| $\beta_2$ | $\tan \beta_2$ | $\sin \beta_2$ | $1 + C_p$ | $u_2$ m/s 1000 rpm | $u_2$ m/s 1100 rpm | $u_2$ m/s 1200 rpm |
|-----------|----------------|----------------|-----------|--------------------|--------------------|--------------------|
| 22 | 0.404 | 0.375 | 0.550 | 17.98 | 18.24 | 18.51 |
| 23 | 0.425 | 0.391 | 0.556 | 17.75 | 18.00 | 18.25 |
| 24 | 0.445 | 0.407 | 0.563 | 17.55 | 17.79 | 18.03 |
| 25 | 0.466 | 0.423 | 0.569 | 17.38 | 17.60 | 17.82 |
| 26 | 0.488 | 0.438 | 0.575 | 17.22 | 17.43 | 17.64 |
| 27 | 0.510 | 0.454 | 0.582 | 17.08 | 17.28 | 17.48 |
| 28 | 0.532 | 0.470 | 0.588 | 16.96 | 17.15 | 17.34 |
| 29 | 0.554 | 0.485 | 0.594 | 16.84 | 17.02 | 17.21 |
| 30 | 0.577 | 0.500 | 0.600 | 16.74 | 16.91 | 17.09 |

$\eta_v$ = volumetric efficiency

$\eta_m$ = mechanical efficiency

Thus, the theoretical head is:

$$H_{th} = H/\eta_h \qquad (6.38)$$

$u_2 = 2u_1$ because initially it was assumed that $(d_1/d_2) = 0.5$. Thus, all terms in Equation (6.36) except $\beta_2$ are known.

$\beta_2$ has been evaluated for a range of N. The values are presented in Table 6-3.

The values of Table 6-3 may now be compared with the values of Table 6-2. We need to match the values of $u_2$ in each table. We see that at a value of 1100 rpm in Table 6-1, $u_2 = 17.28$ m/s and $\beta_1^1 = 23.5°$.

The inlet angle is further adjusted by means of the equation:

$$\tan \beta_1^{11} = \tan (23.5)[\cos (27)] = 0.3874 \tag{6.39}$$

$\therefore \beta_1^{11} = 21.2°$. This is the starting angle in Table 6-4 at $r = 0.075$ m.

There is a match in Table 6-3 for 1100 rpm of $u_2 = 17.28$ m/s at a value of $\beta_2 = 27°$. The assumed blade number of 8 is checked by using the Pfleiderer equation:

$$z = 6.5[(d_2 + d_1)/(d_2 - d_1)] \sin (\beta_1 + \beta_2)/2 \tag{6.40}$$

Substituting values:

$$z = 6.5[(1 + 0.5)/(1 - 0.5)] \sin (20.5 + 27)/2 = 7.9 \tag{6.41}$$

This value of z is close enough to the assumed value of 8 that no further changes are necessary.

Definitions:

1. Pitch is defined as: $t_1 = (\pi d_1)/(z)$.
2. su1 is defined as: $su1 = $ blade thickness$/\sin \beta_1'$.
3. Inlet constriction coefficient $\Phi_{INLET}$ is defined as: $\Phi_{INLET} = t_1/(t_1 - su1)$.

$$t_1 = (\pi d_1)/(z) = (\pi)(0.15)/8 = 0.058 \text{ m or } 5.8 \text{ cm} \tag{6.42}$$

Assuming a blade thickness of 5 mm (0.005 m):

$$su1 = (0.005)/\sin(23.5) = 0.0125 \text{ m or } 1.25 \text{ cm} \tag{6.43}$$

$$\Phi_{INLET} = t_1/(t_1 - su1) = (5.8)/(5.8 - 1.25) = 1.27 \tag{6.44}$$

Area of blade inlet section:

$$A_1 = \Phi_{INLET}(Q^1/c_{m1}) = (1.27)(0.222/3.235) = 0.087 \text{ m}^2 \tag{6.45}$$

Width of impeller at inlet:

$$b_1 = A_1/(d_1\pi) = 0.087/(0.15\pi) = 0.185 \text{ m} = 18.5 \text{ cm} \tag{6.46}$$

$$t_2 = (\pi d_2)/(z) = (\pi)(0.30)/8 = 0.116 \text{ m or } 11.6 \text{ cm} \tag{6.47}$$

$$su2 = (0.005)/\sin (27.0) = 0.0110 \text{ m or } 1.10 \text{ cm} \tag{6.48}$$

$$\Phi_{OUTLET} = t_2/(t_2 - su2) = (11.6)/(11.6 - 1.10) = 1.10 \tag{6.49}$$

Area of blade outlet section:

$$A_2 = \Phi_{\text{OUTLET}} \, (Q^1/c_{m2}) = (1.10)(0.222/3.859) = 0.063 \, \text{m}^2 \tag{6.50}$$

Width of blade outlet section:

$$b_2 = A_2/(d_2 \, \pi) = (0.063)/(0.30 \, \pi) = 0.067 \, \text{m} = 6.7 \, \text{cm} \tag{6.51}$$

Summary of calculations for impeller dimensions

1. Using a range of impeller speeds of 1000 to 2000 rpm, the specific speeds were calculated using Table 6-1.
2. Using the equations of Table 6-1, values of $K_{cm1}$ and $K_{cm2}$ were calculated.
3. Equation (6.24) was used to tabulate values of $c_{m1}$.
4. Values of $d_{\text{shaft}}$ were found using Equation (6.26) and $d_{\text{hub}}$ values from Equation (6.28).
5. Using a volumetric efficiency of 0.96, values of $Q^1$ were found from Equation (6.29). For this example $Q^1 = 0.208 \, \text{m}^3/\text{s}$.
6. $A_0^1$ values were tabulated using Equation (6.31) and Table 6-2.
7. Using the remaining equations, an iterative solution must be sought because the variables are interdependent. Because the number of calculations may be extensive, it is easier to use a spreadsheet to perform them.

Blade shapes were determined by the point-by-point method (Pfleiderer). A linear variation of $c_m$ was assumed, and the variation of w was calculated on the basis of the segment that sloped down. Integration between $r_1$ and $r_2$ was carried out with the equation:

$$\theta = (180/\pi) \int_{r1}^{r2} (r \tan \beta)^{-1} dr \tag{6.52}$$

Numerical integration was carried out between the limits $r_1$ and $r_2$ for finite increments of $\Delta r$. It was assumed that the blades were of constant thickness 5 mm in the radial direction. The values are given in Table 6-4.

Figure 6-5 shows a schematic of the designed impeller, and Figure 6.6 shows a plan view of the impeller blades in accordance with Table 6-4.

## Comparison with Values Predicted by Other Equations

The most important prediction for a given set of initial parameters is that of the tangential component of velocity, $u_2$, at outlet. This may be calculated from the slip factor and determines the performance of the pump.

**Table 6-4**

| Point | R, m | Δr | $c_m$ | w m/s | $\sin\beta = c_m/w$ | β | $1/r\tan\beta$ | Δa | Σ(Δa) | θ |
|---|---|---|---|---|---|---|---|---|---|---|
| 1 | 0.0750 | — | 3.235 | 8.946 | 0.3616 | 21.2 | 34.38 | 0.000 | 0.000 | 0.0 |
| 2 | 0.0825 | 0.0075 | 3.297 | 8.825 | 0.3704 | 21.7 | 30.10 | 0.403 | 0.403 | 23.1 |
| 3 | 0.0900 | 0.0075 | 3.360 | 8.775 | 0.3794 | 22.3 | 26.81 | 0.356 | 0.759 | 43.5 |
| 4 | 0.0975 | 0.0075 | 3.422 | 8.740 | 0.3883 | 22.8 | 24.10 | 0.318 | 1.077 | 61.7 |
| 5 | 0.1050 | 0.0075 | 3.485 | 8.700 | 0.3979 | 23.4 | 21.78 | 0.287 | 1.364 | 78.1 |
| 6 | 0.1125 | 0.0075 | 3.547 | 8.660 | 0.4066 | 24.0 | 19.80 | 0.260 | 1.624 | 93.0 |
| 8 | 0.1200 | 0.0075 | 3.609 | 8.625 | 0.4159 | 24.6 | 18.09 | 0.237 | 1.861 | 106.6 |
| 9 | 0.1275 | 0.0075 | 3.672 | 8.590 | 0.4253 | 25.2 | 16.59 | 0.217 | 2.078 | 119.0 |
| 10 | 0.1350 | 0.0075 | 3.734 | 8.560 | 0.4347 | 25.8 | 15.28 | 0.199 | 2.277 | 130.4 |
| 11 | 0.1425 | 0.0075 | 3.797 | 8.53 | 0.4444 | 26.4 | 14.12 | 0.184 | 2.461 | 140.9 |
| 12 | 0.1500 | 0.0075 | 3.859 | 8.500 | 0.4540 | 27 | 13.08 | 0.170 | 2.631 | 150.7 |

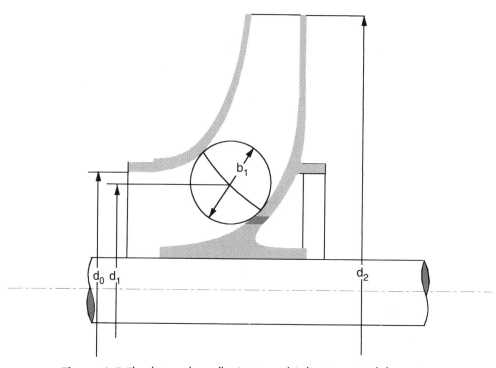

**Figure 6-5** The designed impeller (not to scale) showing critical dimensions.

The Pfleiderer slip factor is:

$$\mu = 1/\{1 + (0.85/z)(1 + \beta_2/60)[2/(1 - r_1^2/r_2^2)]\} \qquad (6.53)$$

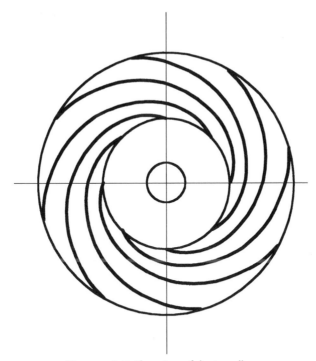

**Figure 6-6** Plan view of the impeller.

Using the data generated:

$$\beta_2 = 27.0°; \ N = 1100 \text{ rpm}; \ (d_1/d_2) = 0.5 \text{ and } u_2 = 17.28 \text{ m/s}$$

The Pfleiderer slip factor is 0.83.

The Busemann slip factor, using the Busemann equation and Figure 4-24, is:

$$\sigma_B = (0.8 - \Phi_2 \tan \beta_2)/(1 - \Phi_2 \tan \beta_2)$$

where:

$$\Phi_2 = (c_{r2}/u_2)$$

Using the same data again, the Busemann slip factor is 0.78.

The tangential component of velocity at outlet $c_{\theta 2} = u_2 - c_{m2} \cot \beta_2 = 9.71$ m/s

Allowing for slip:

$$c_{\theta 2}^1 \text{ (Pfleiderer)} = (9.71)(0.83) = 8.06 \text{ m/s}$$

$$c_{\theta 2}^1 \text{ (Busemann)} = (9.71)(0.78) = 7.57 \text{ m/s}$$

The difference between the two values is the same as the difference between the slip factors: that is, ~6%. In fact, it is generally agreed that the Pfleiderer slip factor equation gives values that are somewhat higher than the true values, so that the agreement is even better.

# 6.4 Design of Blades of Double Curvature

The techniques used in the design of impellers and runners of single curvature may also be used for double-curvature blade design. The methods are applicable to radial flow and diagonal impellers. For pumps care must be taken with regard to the choice of blade length. Too short a length has an unfavorable influence on suction capacity and efficiency. If the velocity of the liquid before the blade and the constriction coefficient are constant, $c_{m1}$ will be the same for all points along the blade. Therefore, the blade inlet angle $\beta_1$ must be variable along the edge; that is, the blade must be twisted or have double curvature.

In certain cases it may be necessary to design a purely radial blade with double curvature because the inlet is wide; these are usually low-specific speed impellers. Wide inlets can mean large variation of $c_{m1}$ along the inlet edge of the blade. Boiler feed pumps, which require a stable H-Q characteristic, also have extended edges; these also should be designed with double curvature.

## 6.4.1 Impeller Blades with Double Curvature

### Procedure

The following procedure may be used; it is similar to that for single-curvature blades:

1. A value of Ns for given values of Q and H is calculated, $u_2$ is calculated, impeller diameter $d_2$ for the central streamline.
2. $c_{m2}$ is found, $\beta_2$ is assumed, and impeller width $b_2$ is calculated.
3. The impeller profile is provisionally assumed together with the position of inlet edge, making sure that the shape is smooth and continuous and that the change from $c_{m1}$ to $c_{m2}$ is gradual.
4. Corrections are made to the profile and position of inlet edge if necessary. For instance, referring to Figure 6-7, streamline $B_1$-$B_2$, the inner shroud streamline may be too short relative to $A_1$-$A_2$, the center streamline, and $C_1$-$C_2$, the inner shroud streamline. It may be lengthened by moving the point $B_1$ in the direction of the inlet or by alternatively shortening $C_1$-$C_2$ by moving $C_1$ in the direction of $C_2$.

### Constriction Coefficient

Referring to Figure 6-7, the following relations are relevant (with the symbols the same as for single-curvature design):

$$s_1^1 = s_1/\sin \lambda_1 \tag{6.54}$$

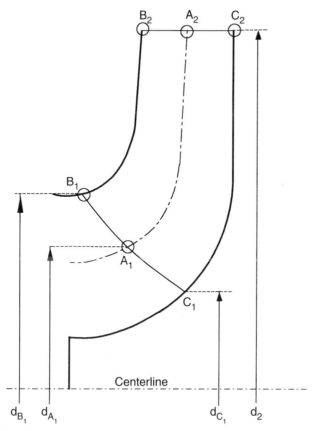

**Figure 6-7** Impeller with extended inlet with three streamlines: $B_1$–$B_2$-inner shroud streamline; $A_1$–$A_2$-center streamline; $C_1$–$C_2$-inner shroud streamline.

$\lambda_1$—the angle between the inlet edge and the streamline at the inlet—is calculated from the relation:

$$\cot \lambda_1 = \cot \lambda_1^1 \cos \beta_2 \tag{6.55}$$

$$s_{u1} = s_1^1/\sin \beta_2 = s_1/(\sin \beta_2 \sin \lambda_1) \tag{6.56}$$

These relations are based on the assumption that the inlet edge lies on the plane of the impeller axis. This is of course not true, but the assumption has a negligible effect on calculation accuracy. The inlet constriction coefficient is given by:

$$1/\Phi_1 = 1 - s_{u1}/t_1 = 1 - (s_1/t_1)[1/(\sin \beta_2 \sin \lambda_1)] \tag{6.57}$$

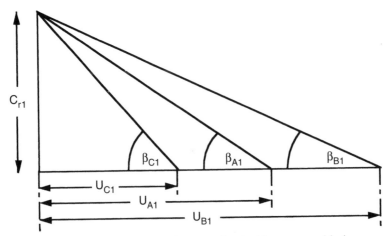

**Figure 6-8** Inlet velocity diagrams for double-curvature blade.

Thus, the final equation defining the contraction coefficient is:

$$1/\Phi_1 = 1 - s_{u1}/t_1 = 1 - (s_1/t_1)[1 + (\cot^2\beta_2 \sin^2\lambda_1)] \tag{6.58}$$

### Inlet Angle, $\beta_1$

This angle varies along the inlet edge. In order to determine the blade shape, the impeller is divided into a number of streams. The streamlines are drawn so that velocity potential lines are orthogonal, remembering that the inner and outer shroud boundaries are also streamlines. Circles are inscribed tangential to segments of the trajectories. Velocity $c_m$ maintains a constant value along a trajectory. When the streamlines are correct, the following relations hold:

$$2r_1^1 \pi d_1^1 = 2r_1^{11} \pi d_1^{11}: 2r_{11}^1 \pi d_{11}^1 = 2r_{11}^{11} \pi d_{11}^{11}: \text{etc.,} \quad \text{or}$$

$$r_1^1 d_1^1 = r_1^{11} d_1^{11}: r_{11}^1 d_{11}^1 = r_{11}^{11} d_{11}^{11}: \text{etc.} \tag{6.59}$$

Figure 6-8 shows inlet velocity diagrams for a double-curvature blade and Figure 6-9 shows streamline construction in an impeller channel.

The blade inclination angles $\beta_1$ are calculated from the central streamline equation:

$$\tan \beta_{1A} = c_{m1}/u_{1A} \tag{6.60}$$

As with single-curvature design, the angle should be increased by the incidence value $\delta_1 = 2\text{--}6°$. That is,

$$\beta_{1A}^1 = \beta_{1A} + \delta_1 \tag{6.61}$$

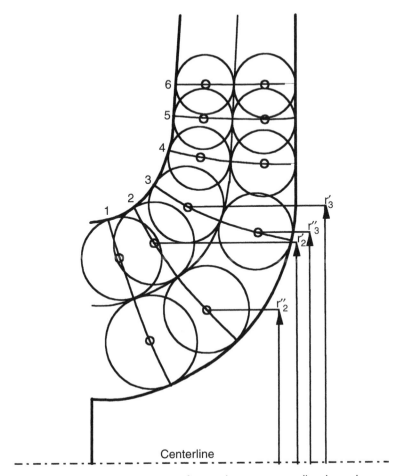

**Figure 6-9** Constructions of streamlines in an impeller channel.

The other blade angles for the other streamlines are determined from the relations:

$$\tan \beta_{1B}^1 = \tan \beta_{1A}^1 (u_{1A}/u_{1B}) = \tan \beta_{1A}^1 (d_{1A}/d_{1B}) \qquad (6.62)$$

$$\tan \beta_{1C}^1 = \tan \beta_{1A}^1 (u_{1A}/u_{1C}) = \tan \beta_{1A}^1 (d_{1A}/d_{1C}) \qquad (6.63)$$

## Determination of $d_2$ and Outlet Angle, $\beta_2$

If the $N_S$ of the impeller is less than 70, the outlet edge of the impeller may remain parallel to the axis of rotation. Exceptions are impellers with diagonal blades; these edges are usually oblique even at low specific speeds. For impellers with a parallel edge, the method of calculation is similar to that of single-curvature impellers.

The optimal number of blades of double curvature is calculated from:

$$z = 13(r_m/e)\sin[(\beta_1 + \beta_2)] \qquad (6.64)$$

where:

$r_m = (M_{st}/e)$: $M_{st} = $ static moment: $r_m = (r_1 + r_2)/2$: $e = r_2 - r_1$

The correction factor $C_p$ is given by:

$$C_p = (\psi^1 r_2)/(z M_{st}) \qquad (6.65)$$

When $(d_2/d_0) < 1.9$:

$$\psi^1 = (1 \text{ to } 1.2) \times (1 + \sin \beta_2)(r_1/r_2) \qquad (6.66)$$

When $(d_2/d_0) > 1.9$:

$$\psi^1 = (0.55 - 0.68) + 0.6 \sin \beta_2 \qquad (6.67)$$

# 6.5 Design of Double-curvature Blades by Conformal Mapping

V. Kaplan, the turbine designer, presented a variation of conformal mapping (see Chapter 2) that gives rapid results. Figures 6-10 and 6-11 best illustrate the method.
The basic construction is as follows:

1. The back and front shroud streamlines are divided into several segments. In Figure 6-10, five segments have been arbitrarily chosen: $p_1$, $p_2$, . . . etc. For greater accuracy, 10 to 15 segments would be preferable.
2. Planes are drawn through the division points $C_1 : J : K : L : C_2$ perpendicular to the impeller axis. The traces of the intersection with the surface of the back shroud are concentric circles—$q_1 : q_2 : q_3$ . . . (Figure 6-11).
3. The segments $e_1 : e_2 : e_3 : e_4$ and $f_1 : f_2 : f_3 : f_4$ and $g_1 : g_2 : g_3 : g_4$ form curvilinear triangles.
4. The radii of the points $C_1 : J : K : L : C_2$ are determined as in Figure 6-10.
5. From the number of blades and the angle of overlap (usually $35°$–$50°$) the central angle $\phi$ may be determined (Figure 6-9).
6. If the outside edge is oblique, then the streamlines will be separated accordingly.

From the hydraulic point of view it is preferable that the angles between the impeller and shrouds be as close to $90°$ as possible. If the shrouds are highly curved, this is almost impossible to manage.

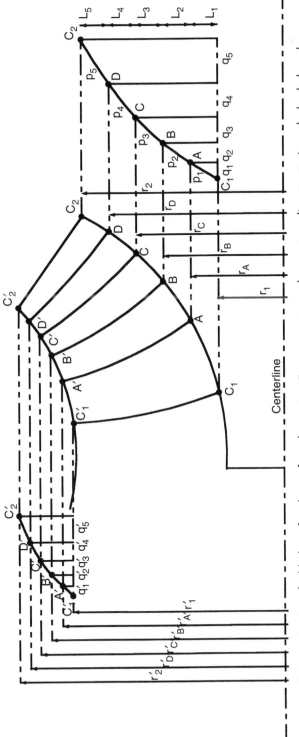

**Figure 6-10** Determination of a blade surface by conformal mapping. Five segments are shown: streamline $C_1$–$C_2$ is on the back shroud, and streamline $C_1^1$–$C_2^1$ is on the front shroud.

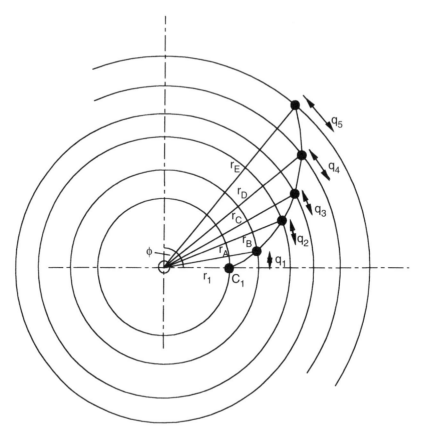

**Figure 6-11** Determination of a blade surface by conformal mapping. End elevation projection of the back shroud showing curvature. Central angle determination Φ is shown.

# 6.6 References

Pfleiderer, C., *Strömungsmaschinen*, Springer-Verlag, Berlin (1957).

Stepanoff, A.J., *Centrifugal and Axial Flow Pumps. Theory, Design and Application* (2nd Ed.). John Wiley & Sons., New York (1957).

# INLET AND OUTLET ELEMENTS

## 7.1 Inlet Elements of Turbines

The inlet elements of turbines such as penstocks and head level control have effectively been covered in Chapter 3. The actual way in which water is introduced into guide vanes is a function of the head and the volumetric flow rate of the water available. For example, we may compare Figures 3-3, 3-17, and 3-18 in Chapter 3.

One aspect that is of importance because of the large surge and flood volumes that may be associated with large turbine systems is the control of inflow and outflow. Inflow surge may cause serious water hammer problems at stay vanes and guide vanes.

Synchronous bypassing of excess water is a desirable solution to this problem, but slow gate response for the bypass is also a problem for rapid flow rate changes. The main difficulty with this solution is that the design of such a bypass would have to allow for the possible high maximum value of such excess water; that is, a large infrastructure would be required. Such a solution would be expensive. An alternative solution that is both effective and less expensive is to provide a surge tank. Similar to a water hammer itself, the fluid phenomenon occurring in a surge tank is periodic. The difference is that the period of fluid oscillation is low. In large surge tanks, this period may vary from several minutes to several tens of minutes. In comparison, water hammer damping may take from several seconds to a few tens of seconds. The reason for the large difference is that the mass of water in the surge tank is so large that the acceleration of it is small. Fortunately, although both phenomena are obviously connected, because one is a consequence of the other, the periodicity difference enables them to be treated separately.

### 7.1.1 Surge Tanks

In its simplest form, a surge tank is a tank connected to a flow line through which water may be in a transient condition. The tank may be connected to an orifice or a compressed air regulator. A surge tank invented by Johnson (1915) in which the water storage function is separated from the acceleration function has a rapid response time. This type of surge tank is called a differential

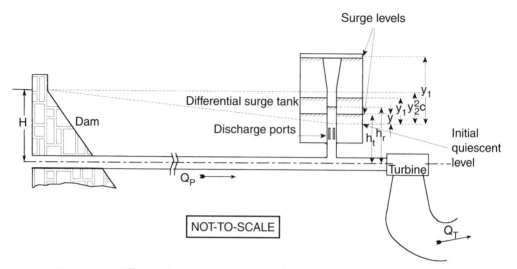

**Figure 7-1** Differential surge tank upstream of a turbine. Load rejection condition.

surge tank. Figure 7-1 shows such a tank schematically. In the load rejection mode as shown, the water level in the tank is at its highest; in the load demand mode, it is lower.

### 7.1.2 Basic Equations for Differential Surge Tanks

At any instant, the following equations hold for the system:

**Load demand:**

$$dt = [(L/g)\,dv]/[y_1 - C\,(v^2 - v_1^2)] \tag{7.1}$$

$$A_P\,v_2\,dt = A_{NET}\,dy + Av\,dt \tag{7.2}$$

**Load rejection:**

$$dt = [(L/g)\,dv]/[y_1 - C\,(v_2^2 - v^2)] \tag{7.3}$$

$$A_P\,v_1\,dt = A_{NET}\,dy + Av\,dt \tag{7.4}$$

where:

$A_P$ = penstock area
$A_{NET}$ = net area of tank = total tank area − riser area
$C$ = coefficient for head losses in penstock
$H$ = reservoir head at conduit
$L$ = penstock length

$v$ = velocity in penstock at any instant

$v_1$ = initial velocity in penstock before acceleration begins

$v_2$ = initial velocity in penstock before deceleration begins

Referring to Figure 7-1, in the rejection condition, there are two cases to be considered, depending on whether or not the surge rises above the reservoir level.

In the first case, if $y_1 < Cv_2^2$, the equation for $A_{NET}$ is:

$$A_{NET} = (A_P L/2gCy_1)[\ln(v_2^2 - P_1^2)/(v_1^2 - P_1^2)$$
$$-(v_1/P_1)\ln[(v_2 - P_1)(v_1 + P_1)]/[(v_2 + P_1)(v_1 - P_1)]] \quad (7.5)$$

In the second case, if $y_1 > Cv_2^2$, the equation for $A_{NET}$ is:

$$A_{NET} = (A_P L/2gCy_1)[\ln(v_2^2 + P_0^2)/(v_1^2 + P_0^2) - (2v_1/P_0)[\arctan(v_2/P_0) - \arctan(v_1/P_0)]] \quad (7.6)$$

where:

$P_1 = (v_2^2 - y_1/C)^{1/2}$
$P_0 = (y_1/C - v_2^2)^{1/2}$

In practice, the most common case of interest is where there is complete shutdown with $v_1 = 0$. $A_{NET}$ then becomes:

$$A_{NET} = (A_P L/2gCy_1)[\ln\{[(v_2^2 + P_0^2)/P_0^2]\}] \quad (7.7)$$

### 7.1.3 Instability of the Surge Tank

A surge tank is able to dampen oscillations following any load changes. For large load changes, the water velocity through the ports is also large, with a marked difference between levels in the tank and the riser. This in turn throttles and suppresses any transience or periodicity. The tank diameter is critical. A larger than necessary tank means that penstock friction will quickly dampen any surge from small load changes. On the other hand, if the tank is too small, water levels in the tank will rise rapidly, leaving insufficient time for penstock friction to be fully effective. As a result, changes in water level over time could lead to magnification of small load variations.

## 7.2 Inlet Elements of Pumps

The best elements are those that have cross-sectional changes that are smooth and gradual, providing uniform changes in flow. The main types are:

1. Straight entry.
2. Bell mouths. These are used for large volumetric flow rate, diagonal-flow pumps.

3. Bends with a large radius of curvature. Bends are not suitable for high $N_S$ pumps.
4. Uniform straight or tapered straight. These should have flat tops to avoid air pockets and should be used in horizontal pumps.
5. Concentric suction chambers. These are used primarily for multistage pumps.
6. Volute type. These are fairly common and are used in single- and double-entry pumps.

Figure 7-2 shows a pump with a straight-pipe fitting. Inset diagrams (a)–(c) show the types of inlet fittings 2 to 4 above, together with a typical pump onto which they would be fitted. Section A on the inlet side would be where these are located.

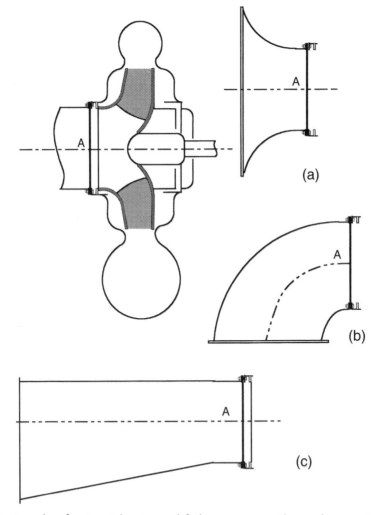

**Figure 7-2** Examples of various inlets. Upper left shows a pump with straight entry. (a) Bell mouth; (b) wide-bend entry; (c) horizontal tapered, flat top entry—avoiding air bubbles.

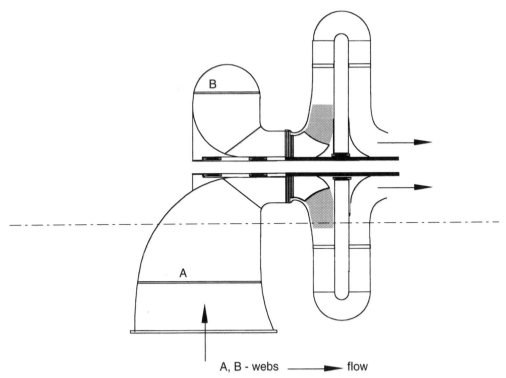

A, B - webs ⟶ flow

**Figure 7-3** First stage of a multistage pump showing a concentric suction chamber fitted with internal webs.

Figure 7-3 shows the first stage of a multistage pump. The flow exiting from the first stage shown would enter an identical second stage; the exit from the second stage would enter the third stage and so on. The internal webs labeled A and B prevent internal circulation in the inlet chamber, so that the flow into the suction side is uniform and energy is not lost in circulating flow that otherwise would be contained in the main body of the flow.

Figure 7-4 shows a double-entry pump. In effect, this is the pump of Figure 7-2 with its mirror image about the centerline joined at the vertical axis of symmetry. There are webs on both sides similar to the pump of Figure 7-3.

## 7.3 Outlet Elements of Turbines

### 7.3.1 Draft Tubes

The conduit from the exit of the runner to the tailrace constitutes the draft tube of a reaction turbine. Runners of reaction turbines have no contact with atmosphere and may be positioned

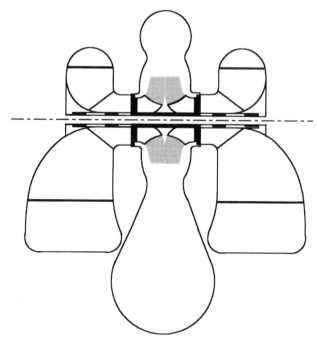

**Figure 7-4** A typical double-entry pump with volute suction chambers.

higher than the tailwater without losing head. The difference in head between the runner exit and tailwater level is designated as the suction head, $H_S$ (see Figure 7-5). When conditions are static, the absolute pressure at the runner exit is less than atmospheric by the amount $H_S$. This implies, of course, that the draft tube is sealed so that it is air tight.

The primary purpose of any draft tube is to recover as much of the velocity head at runner exit as possible. Energy recovery is usually expressed in the form of a head recovery $\Delta H$:

$$\Delta H = \eta_D(c_2^2/2g) \tag{7.8}$$

where:

$\eta_D$ = draft tube efficiency
$c_2$ = absolute velocity at runner exit

It is noted that $(c_2^2/2g)$ is a function of specific speed. One implication of a high value of $\eta_D$ is that the value of $ND_2^3/Q_1$ may be smaller; that is, this fixes the value of $D_2$.

There are limitations on the position, size, and shape of the draft tube. Cavitation limitations, for example, restrict the position of the turbine to the point where it cannot be placed higher than

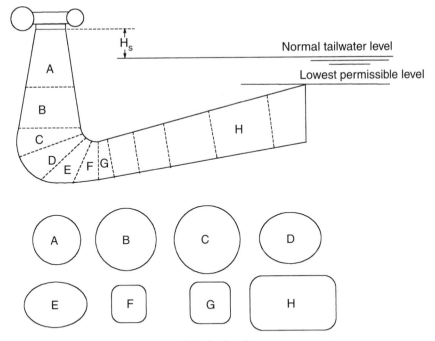

**Figure 7-5** Typical draft tube showing cross sections.

the tailwater level given by:

$$H_S \# H_{ATM} - \sigma_C H \tag{7.9}$$

Typical shape changes along a draft tube are shown in Figure 7-5. The vertical portion of a draft tube has a circular cross section; thereafter the shape changes after the bend joining the inclined or horizontal section. The amount of excavation and associated civil engineering costs for the draft tube should also be kept to a minimum. Thus, it is usually not possible to use a long, straight, conical-type draft tube for Kaplan and propeller-type turbines because of this. Figures 7-6 and 7-7 illustrate this point.

## 7.4 Outlet Elements of Pumps

Liquids leave the periphery of the impeller at much higher velocities than needed in the delivery pipe. Consequently, the velocity must be reduced in a smooth and shockless fashion, with the exit kinetic energy being transformed into pressure energy. A well-designed outlet element does this in as efficient a way as possible through the following methods:

1. Vortex chamber
2. Concentric casing

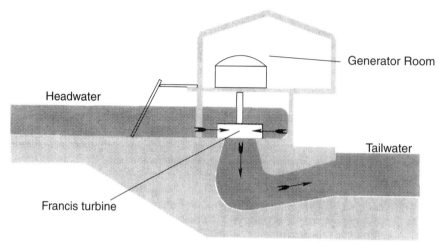

**Figure 7-6** Typical Francis turbine draft tube installation.

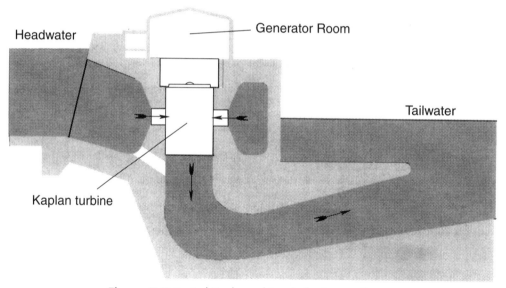

**Figure 7-7** Typical Kaplan turbine draft tube installation.

3. Volute casing
4. Various forms of diffuser vanes

Because volutes are the most common way of achieving the above, we shall concentrate on these and their design.

## 7.4.1 Volute Design

Volutes are designed on the basis of two different principles:

1. Constant moment of momentum
2. Constant mean velocity

A volute design commonly used is in the form of an Archimedean spiral, the construction of which is shown in Figure 7-8. An Archimedean spiral is defined by the condition that its distance from the center of the system changes by equal steps, r, for equal angular steps, ϕ. The curve between intersecting points may be approximated by a circular arc as shown. The shape of the outer contour of a volute casing is the most common application of this spiral. In such a construction, increments in cross-sectional area, corresponding to constant angular steps, are proportional to the distance of the additional area from the center of the system.

The layout shown in Figure 7-8 between 1 and 8 follows this description except for the radial sections where the spiral is pushed out to make room for the casing tongue, thickness t. Typical volutes of circular cross section look like this.

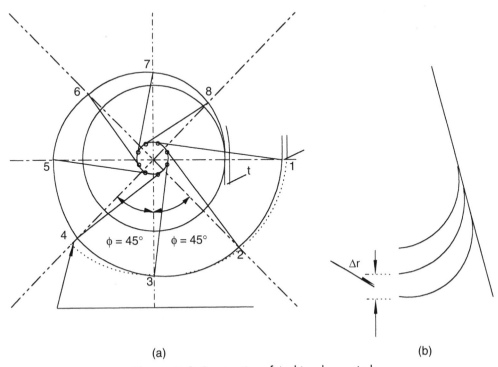

(a)                                                  (b)

**Figure 7-8** Construction of Archimedean spiral.

For a volute of arbitrary cross section, the elemental volumetric amount of fluid dQ flowing through an elemental area da (at a right angle to the radius) is:

$$dQ = da \ c_u = (bdrM_m)/r \tag{7.10}$$

$da = (bdr)$ and $M_m$ is the moment of momentum.

The volume flowing through any cross section bounded by the radii $r_A$ and $r_B$ is:

$$Q_S = (bdrM_m)/r \tag{7.11}$$

If $Q_S$ is $(\phi°/360°)$ of the whole discharge, that is, $Q_S = (\phi°/360°)Q$ then:

$$\phi° = 360 \, M_m/Q \int (bdr)/r \tag{7.12}$$

The elemental area da is a small trapezium. For a volute of circular cross section, the same relationships hold except that we have by simple trigonometry:

$$(b/2)^2 + (r - a)^2 = \rho^2 \tag{7.13}$$

Substitution in Equation (7.12) yields:

$$\phi° = (720\pi M_m/Q)[a - (a^2 - \rho^2)^{1/2}] \tag{7.14}$$

Pfleiderer (1957) presents an empirical equation that allows for friction:

$$\Delta\rho = 0.025 r_A (\phi°/360°) \tag{7.15}$$

$\Delta\rho$ represents the allowance in terms of an increase in radius $\rho$ to $\rho + \Delta\rho$. It should be noted that Equation (7.15) assumes a value of friction factor of 0.0475. If the surface of the volute were different, it would have to be corrected accordingly.

Stepanoff has also presented a plot; see Figure 7-9 relating the average volute velocity, $c_3$, given by:

$$c_3 = K_3(2gH)^{1/2} \tag{7.16}$$

to the specific speed $N_S$. Note that the volute angle in this plot is equivalent to the central angle $\phi$ in the notation of the design example of Section 6.

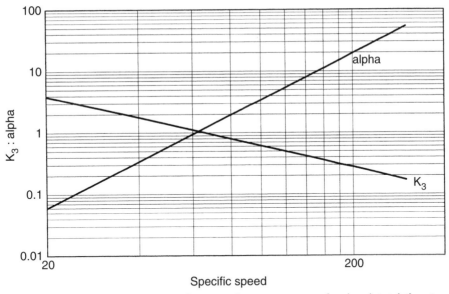

**Figure 7-9** Volute velocity constant as a function of $N_S$. From: *Centrifugal and Axial Flow Pumps* by A.J. Stepanoff. Copyright ©1957 by John Wiley & Sons, Inc. This material is used by permission of John Wiley & Sons, Inc.

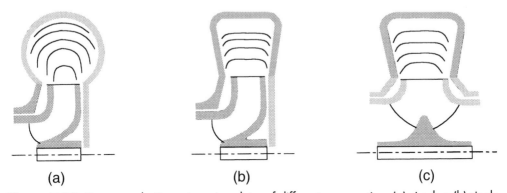

**Figure 7-10** Constant-velocity contours in volutes of different cross section: (a) circular; (b) single-entry, flattened volute; (c) double-entry flattened volute.

## 7.4.2 Velocity Distributions in Different Volute Cross Sections

Figure 7-10 shows three different volute cross sections that are typically used with the constant velocity contours (isovels) in the volutes.

### 7.4.3 Design of a Volute

The volute converts the velocity head to pressure head as efficiently as possible. Once the general shape of the volute has been decided on, such as one of those in Figure 7-10, the design method for any volute is the same. The flow in the volute is close to a free-vortex flow.

Referring to Figures 7-11(a), (b), and (c) as a typical volute shape, we may write the differential flow rate $dQ_\phi$ through an elemental area $dA$ as:

$$dQ_\phi = dA \, c_u = b \, dR \, c_u \tag{7.17}$$

The equation of a free vortex is:

$$R \, c_u = a \text{ constant} = C \tag{7.18}$$

Therefore, the total flow through any cross section is:

$$Q_\phi = \int_{R2}^{Rphi} dQ = C \int_{R2}^{Rphi} (b/R) \, dR \tag{7.19}$$

The volute section in Figure 7-11(b) is approximated by the trapezium of Figure 7-11(c) with negligible loss of accuracy. The area of flow is now the trapezium cross section. The relation between the areas $A_a$ and $A_b$ is given by:

$$A_a/R_a = A_b/R_b \tag{7.20}$$

where:

$R_a$ and $R_b$ are the radial distances to the centers of gravity of $A_a$ and $A_b$.

Referring to Figure 7-11(c), if $x =$ the width of the cross section at $R_{t\phi}$, then the area of the cross section at any angle $\phi$ is given by:

$$A_\phi = (b_b/2)(R_{t\phi} - R_2) + \tan(\theta/2)(R_{t\phi} - R_2)^2 \tag{7.21}$$

In Equation (7.14) the only variable is $R_{t\phi}$. Therefore, in order to calculate $A_\phi$ as a function of $R$, we need a functional relation between $\phi$ and $R$. The equation for a logarithmic spiral is:

$$R_{t\phi} = R_2 e^{\tan \alpha \, \phi 1} = R_2 e^{\tan \alpha \, \phi \, (\pi/180)} \tag{7.22}$$

where:

$\alpha =$ angle of spiral
$\phi 1 = \phi$ measured in radians

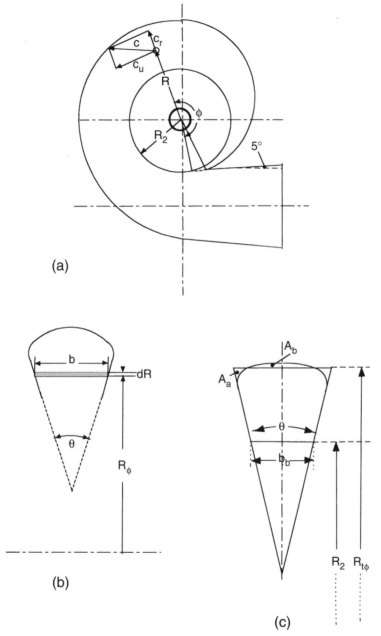

(a)

(b)

(c)

**Figure 7-11** (a) Volute; (b) cross section of volute at any radius; and (c) assumed cross section of volute for calculation.

Equation (7.22) may be more conveniently written as:

$$\phi = [\ln(R_{t\phi_t}/R_2)]/[(0.0175)\tan\alpha] \tag{7.23}$$

Equations (7.21) and (7.22) may now be used in conjunction to calculate the radius and the area of the volute at any radius once a value of $\alpha$ is set for a particular volute. For the tongue radius, at $R = R_t$ the angle corresponding to $\phi = \phi_t$ is:

$$\phi_t = [\ln(R_{t\phi_t}/R_2)]/[(0.0175)\tan\alpha] \tag{7.24}$$

Usually, the angle $\theta$ lies between $40°$ and $60°$, and the diffuser half-angle is about $5°$. The minimum cross section of the volute occurs at $\phi = \phi_t$, and the maximum at $\phi = 360°$.

### 7.4.4 Relation between Volute Velocity and Specific Speed

Stepanoff (1957) has presented a curve relating the velocity coefficient $K_{cv}$ to the specific speed of the pump. The velocity in a volute is given by:

$$c_v = (K_{cv})(2gH)^{1/2} \tag{7.25}$$

The plot of this is shown in Figure 7-12.

## 7.5 Solved Problem

### 7.5.1 Pressure Distribution within a Volute

The volumetric flux through a spiral casing, of constant height h, is Q. The fluid is incompressible and inviscid. The volute is designed such that the momentum per unit mass is constant around the circumference. The pressure at A is atmospheric ($p_0$). Calculate the pressure at radius $r_B$.

### Solution
The Bernoulli equation applied to the streamline A-B with no body forces or potential change is:

$$p_0 + (\rho/2)c_A^2 = p_B + (\rho/2)c_B^2 \tag{7.26}$$

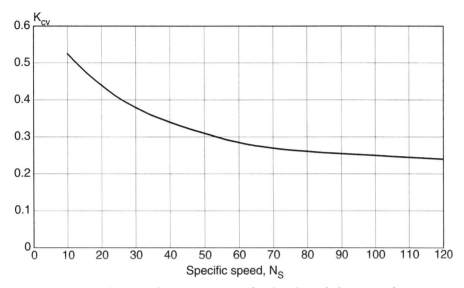

**Figure 7-12** $K_{cv}$ as a function of $N_S$. From: *Centrifugal and Axial Flow Pumps* by A.J. Stepanoff. Copyright ©1957 by John Wiley & Sons, Inc. This material is used by permission of John Wiley & Sons, Inc.

$c^2 = c_r^2 + c\theta^2$ at A and B is unknown

$$c_r = Q/(2\pi rh) = A/r \qquad (7.27)$$
$$A = Q/(2\pi h)$$

Because the fluid is inviscid, $c_2 = K/r$. The constant K is calculated from the condition that the outer casing is a streamline, the equation of which is:

$$(1/r)(dr/d\theta) = c_r/c\theta = -A/K \qquad (7.28)$$

This may be integrated as:

$$\ln r = -(A/K)\theta + \ln C \qquad (7.29)$$
$$r\,(\theta = 0) = R$$

Substituting the boundary condition, Equation (7.27) becomes:

$$\ln(R/r) = (A/K)\theta \qquad (7.30)$$

K is evaluated from: $r\,(\theta = 2\pi) = r_B$; thus $K = (A2\pi)/\ln(R/r_B)$

Finally, $c_A^2 = (A^2 + K^2)/r_A^2$ and $c_B^2 = (A^2 + K^2)/r_B^2$
Substituting in Equation (7.26):

$$p_0 + (\rho/2)(A^2 + K^2)/r_A^2 = p_B + (\rho/2)(A^2 + K^2)/r_B^2 \tag{7.31}$$

$$\therefore \quad p_B - p_A = (\rho/2)(A^2 + K^2)(1/r_A^2 - 1/r_B^2) \tag{7.32}$$

$A = Q/(2\pi h)$ and $K = (A2\pi)/\ln(R/r_B)$

## 7.6 References

Johnson, R.D., "The differential surge tank." *Trans. ASCE* 88, Paper 1324 (1915).

Pfleiderer, C., *Strömungsmaschinen*, Springer-Verlag, Berlin (1957).

Stepanoff, A.J., *Centrifugal and Axial Flow Pumps. Theory, Design and Application* (2nd Ed.). John Wiley & Sons., New York (1957).

# CHAPTER 8

# HEAD LOSSES IN COMPONENTS OF TURBINE AND PUMP SYSTEMS

## 8.1 Pipes

The flows in turbine/pump systems are almost invariably highly turbulent. Complete, theoretical solutions exist for laminar, steady, and unsteady flows in pipes but not for turbulent flows. To predict the behavior of fluids in turbulent flow, either empirical relationships or computer predictions based on turbulent models of fluid behavior must be used. The most convenient way to do this for calculations is to use one of a number of empirical relations that have been developed (Round and Garg, 1986) or, alternatively, to use a friction factor–Reynolds number relation in graphical form, the so-called Moody diagram. The most convenient method for the most practical use is the Moody diagram. However, if the problem requires an iterative method for its solution, the use of an accurate equation, preferably an explicit one in friction factor, should be used. The equation given by Nekrasov (1968) is accurate over a wide range of Reynolds numbers and relative roughness. A number of other implicit equations have been proposed, but they are generally cumbersome to use relative to explicit equations.

### 8.1.1 Friction Factor

Laminar flows in pipes and ducts may be analyzed theoretically, and we begin with laminar flow as a background to a discussion of turbulent flows. The fully developed velocity profile of a Newtonian fluid, water, for example, flowing steadily in a pipe is given by:

$$u = (R^2/4\mu)(-\partial p/\partial x)\,[1 - (r/R)^2] \qquad (8.1)$$

where:

R = pipe radius
r = radius at any point in the flow

$\mu$ = fluid viscosity
x = distance along the pipe
p = pressure

The volumetric flow rate is given by:

$$Q = \int_0^R u\,2\pi r\,dr = (\pi R^4/8\mu)(-\partial p/\partial x) \qquad (8.2)$$

The average velocity is:

$$V_{AV} = Q/(\pi R^2) \qquad (8.3)$$

In fully developed flow in a pipe, the pressure gradient is constant.
$\therefore$ $(-\partial p/\partial x)$ may be written as $-(p_2 - p_1)/(L_2 - L_1) = (\Delta p/\Delta l)$, so that Equation (8.2) becomes:

$$Q = (\pi R^4/8\mu)(\Delta p/\Delta l) = (\pi D^4 \Delta p)/(128\mu\Delta l) \qquad (8.4)$$

If we write:

$$(p_1 - p_2)/\rho g = \Delta p/\rho g = h_f \qquad (8.5)$$

Combining Equations (8.3), (8.4), and (8.5) results in:

$$h_f = 32(\Delta l/D)(\mu/\rho)(V_{AV}/Dg) = (\Delta l/D)(V_{AV}^2/2g)(64\mu/\rho V_{AV}D)$$
$$= (64/Re)(\Delta l/D)(V_{AV}^2/2g). \qquad (8.6)$$

where:

$$Re = (\rho V_{AV}D/\mu).$$

Equation (8.6) is valid regardless of the pipe internal roughness. For turbulent flow this is not true, so that for turbulent flow the pressure drop cannot be evaluated analytically. Insight into the form of the equation must be obtained from experimental data. Using dimensional analysis, we may write the pressure: $\Delta p = -(D, L, e, V_{AV}, \rho, \mu)$. Here e is the pipe roughness. Using dimensional analysis, we may obtain a correlation of the form:

$$\Delta p/(\rho V_{AV}^2) = -[Re, (\Delta l/D), (e/D)] \qquad (8.7)$$

The term on the left-hand side may be more conveniently written as $\Delta p/(\rho V_{AV}^2/2g)$. The units of pressure in this case would be measured in units of height of fluid because the units of $(\rho V_{AV}^2/2g)$

are length. The functional form of the term in square brackets is not known, but from experiments it is known that the left-hand side of Equation (8.7) is directly proportional to $(\Delta l/D)$. Noting that $\Delta p/\rho g = h_f$, Equation (8.7) becomes:

$$h_f/(V_{AV}^2/2g) = (\Delta l/D)\phi[Re, (e/D)] \qquad (8.8)$$

The function on the right-hand side of Equation (8.8) may be replaced by a factor f, the friction factor. So that Equation (8.8) is also commonly written:

$$h_f = f(\Delta l/D)(V_{AV}^2/2g) \qquad (8.9)$$

The friction factor must be determined experimentally. Figure A9-1 (Moody diagram) shows the relationship between friction factor and Reynolds number with relative roughness (e/D) as the other parameter.

### 8.1.2 Hydraulic Diameter

A characteristic dimension, which is useful for noncircular ducts or pipes flowing partially full, is the hydraulic diameter, defined as:

$$D_H = 4(\text{cross-sectional area of flow})/(\text{perimeter in fluid contact}) = 4A/P \qquad (8.10)$$

The definition is based on a circular pipe flowing full. Thus for a circular pipe, $A = (\pi/4)D^2$ and $P = \pi D$.

$$\therefore \qquad D_H = [4(\pi/4)D^2]/(\pi D) = D \qquad (8.11)$$

For a rectangular duct, sides a and b, flowing full:

$$D_H = [4(ab)]/(2a + 2b) = 2ab/(a + b) \qquad (8.12)$$

The hydraulic diameter may be used to define a Reynolds number, $Re = (D_H V)/v$, which in turn may be used with the friction-factor plot to calculate friction factor.

## 8.2 Losses through Other Elements

### 8.2.1 Discharge, Velocity, and Contraction Coefficients

Consider a sharp-edged orifice in the side of a vertical tank as shown in Figure 8-1.

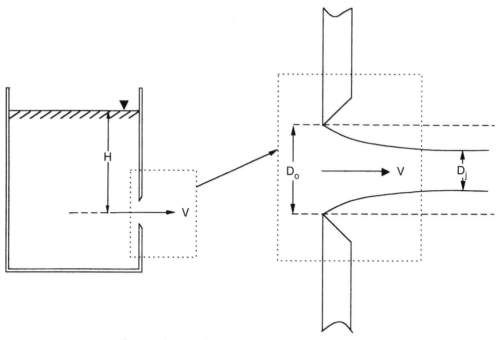

**Figure 8-1** Definition diagram for discharge, velocity, and contraction coefficients.

Application of the Bernoulli equation between points 1 and 2, the top of the fluid in the tank, and the centerline of the outlet gives:

$$H = V^2/2g \quad \text{or} \quad V = (2gH)^{1/2} \tag{8.13}$$

Equation (8.13) applies to an ideal fluid; for a fluid possessing viscosity (i.e., a real fluid), friction reduces the velocity V to a velocity $V_1$, where $V_1$ is given by:

$$V_1 = C_V V = C_V(2gH)^{1/2} \tag{8.14}$$

where $C_V$ is called a velocity coefficient. The jet, after passing through the orifice, contracts to an equilibrium diameter, $D_j$. The ratio $(A_O/A_J)$, where $A_O$ = orifice area and $A_J$ = jet area, is called the contraction coefficient, $C_C$. The theoretical volumetric flow rate is given by:

$$Q_{TH} = A_O(2gH)^{1/2} \tag{8.15}$$

The discharge coefficient for the orifice is defined as:

$$C_D = (Q_{AC}/Q_{TH}) \tag{8.16}$$

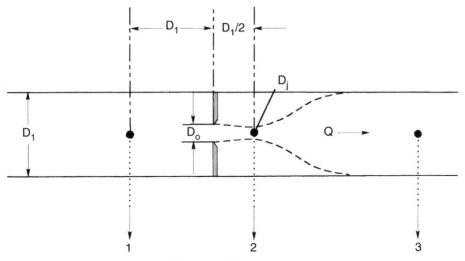

**Figure 8-2** Orifice meter showing pressure tap positions.

where:

$Q_{AC}$ = actual flow rate

$$\therefore \quad Q_{AC} = A_J V_J = C_C C_V A_O (2gH)^{1/2} = C_D A_O (2gH)^{1/2} \quad (8.17)$$

In a similar manner as shown in Figure 8-2, which illustrates the standard pressure tap positions of an orifice meter, the continuity equation gives:

$$A_1 V_1 = A_O V_1 = A_J V_J \quad (8.18)$$

The variation of pressure with position for the meter is shown in Figure 8-3.

Venturi meters also have a representative $C_D$. Consider the venturi meter shown in Figure 8-4. Application of the continuity equation between points 1 and 2 gives:

$$Q_{AC} = A_1 V_1 = A_2 V_2 \quad (8.19)$$

Application of the Bernoulli equation for steady flow between points 1 and 2 gives:

$$(p_1/\rho g) + (V_1^2/2g) + z_1 = (p_2/\rho g) + (V_2^2/2g) + z_2 + h_{12} \quad (8.20)$$

where:

$h_{12}$ = loss of head between 1 and 2

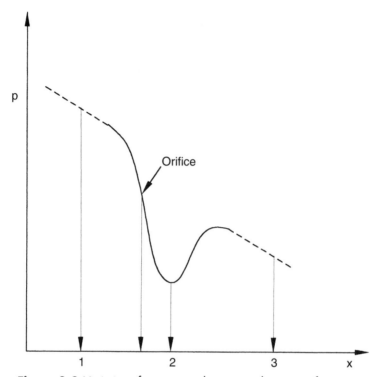

**Figure 8-3** Variation of pressure with position along an orifice meter.

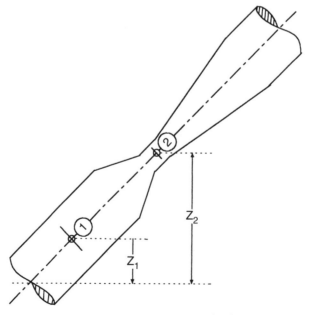

**Figure 8-4** Venturi meter. Minimum pressure occurs at the throat of the meter, position 2.

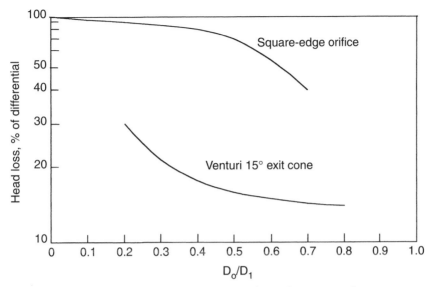

**Figure 8-5** Comparison of head loss across a square-edge orifice meter with a Venturi meter, with an exit cone angle = 15°.

The coefficient of discharge, $C_D$, is defined as in Equation (8.16) and is measured by experiment. A comparison of head loss as a percentage of the total pressure differential across the meter as a function of diameter ratio is shown in Figure 8-5. It can be seen that the head loss across a Venturi meter is always considerably less than that across an orifice meter.

## 8.2.2 Nozzle Loss

The head loss through a nozzle is usually expressed as:

$$H_N = (1/C_V^2 - 1)(V_j^2/2g) \tag{8.21}$$

where:

$C_V$ = velocity coefficient

## 8.2.3 Fittings, Valves, and Joints

Pressure or head losses through fittings in a pipe are customarily expressed as a loss coefficient:

$$\Delta p = -K(\rho V^2/2) \tag{8.22}$$

or

$$\Delta p/\gamma = \Delta H = -K(V^2/2g) \tag{8.23}$$

Loss coefficients for a variety of valves and fittings are shown in Appendix A8.

### 8.2.4 Expansions and Contractions

Figure A8-4 in Appendix A8 shows the loss coefficient for expansions and contractions. It should be noted that velocities to which the coefficients should be applied refer to the smaller diameter pipe.

### 8.2.5 Losses in Pipe Branches

In the same way that the head loss is expressed for fittings of one sort or another, the same can be done for pipe branches. Thus, the head loss is expressed as a coefficient in the form:

$$\Delta H = K(\rho V^2/2g) \tag{8.24}$$

The average velocity, V, is given by the total flow rate divided by the cross-sectional area of the main pipe that is, Q/A.

K is a function of both the flow in the main pipe and in the branch pipe, Q and $Q_B$. K has two sets of values: one for the straight section, $K_S$, and one for the branch section, $K_B$.

Figure 8-6 shows the arrangement and notation for three pipe branches. Figure 8-7 shows the variation of loss coefficients for the three branch pipes of Figure 8-6.

## 8.3 Total Frictional Loss in a Pipe System

Steady-flow head loss for an incompressible, frictionless (ideal) fluid in a piping system may be written:

$$p_1/\gamma + V_1^2/2g + z_1 = p_2/\gamma + V_2^2/2g + z_2 \tag{8.25}$$

Equation (8.25) must be modified to take into account all the losses due to friction and minor losses due to fittings. Thus, two terms are added to the right-hand side of the equation to account for the sum of all the friction losses and the sum of the fitting losses. Equation (8.25) therefore becomes:

$$p_1/\gamma + V_1^2/2g + z_1 = p_2/\gamma + V_2^2/2g + z_2 + \Sigma F(\Delta l/D)(V_{AV}^2/2g) + \Sigma K(V_{AV}^2/2g) \tag{8.26}$$

The $V_{AV}$ term in the summation terms refers to the appropriate velocity at the location considered.

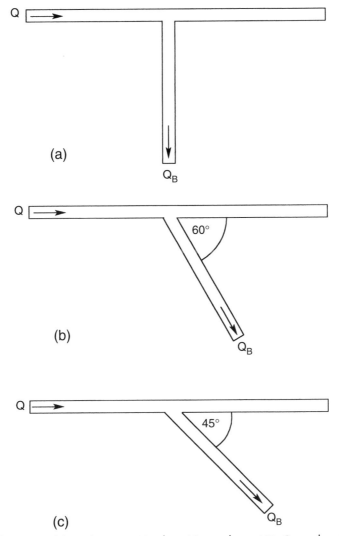

**Figure 8-6** Three typical branches, a = 90°, b = 60°, and c = 45°. Q = volumetric flow through straight section; $Q_B$ = volumetric flow through branch.

## 8.4 Solved Problems

**8.4.1** Three pipes of different diameter are connected in series as shown in Figure 8-8. The data for the pipes are:

1. Length = 250 m, diameter = 30 cm
2. Length = 200 m, diameter = 20 cm
3. Length = 220 m, diameter = 25 cm

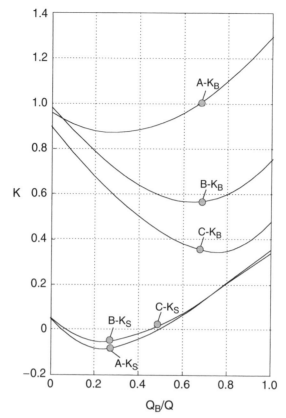

**Figure 8-7** Variation of $K_S$ and $K_B$ as a function of $Q_B/Q$. Subscript S = straight section; Subscript B = branch (A) 90° branch; (B) 60° branch; (C) 45° branch.

Inlet pressure is 300 kPa(g), and discharge is to atmosphere, 101 kPa(a). The roughness of each pipe, $\varepsilon$, is 0.005 cm. What is the discharge flow rate?

## Solution

Relative roughness of each pipe:

1. $(\varepsilon/D_1) = (0.005/0.30) = 1.67 \times 10^{-2}$
2. $(\varepsilon/D_2) = (0.005/0.20) = 2.50 \times 10^{-2}$
3. $(\varepsilon/D_3) = (0.005/0.25) = 2.00 \times 10^{-2}$

The relevant diameter ratios are:

$$D_2/D_1 = 0.667$$

$$D_2/D_3 = 0.80$$

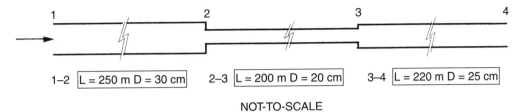

1-2  L = 250 m D = 30 cm    2-3  L = 200 m D = 20 cm    3-4  L = 220 m D = 25 cm

NOT-TO-SCALE

**Figure 8-8** Pipes of different diameter connected in series.

Referring to Figure A8-4, the resistance coefficient for the contraction = 0.24. The resistance coefficient for the expansion = $[(1 - (D_2/D_3)^2]^2 = 0.13$. The values of the resistance coefficients are based on velocities in the smaller diameter pipe.

Referring to Figure 8-8, the total head loss across the system is:

$$H = h_{f1\text{-}2} + h_2 + h_{f2\text{-}3} + h_3 + h_{f3\text{-}4} + h_4 \qquad (8.27)$$

where:

$$h_{f1\text{-}2} = f_1(L_1/D_1)(V_1^2/2g) \qquad (8.28)$$

$$h_2 = 0.24(V_2^2/2g) \qquad (8.29)$$

$$h_{f2\text{-}3} = f_2(L_2/D_2)(V_2^2/2g) \qquad (8.30)$$

$$h_3 = 0.13(V_2^2/2g) \qquad (8.31)$$

$$h_{f3\text{-}4} = f_3(L_3/D_3)(V_3^2/2g) \qquad (8.32)$$

$$h_4 = (V_3^2/2g) \qquad (8.33)$$

The velocities in each pipe are related to each other as:

$$(V_2/V_1) = (D_1/D_2)^2 = 2.25; \quad V_2^2 = 5.06V_1^2 \qquad (8.34)$$

$$(V_3/V_1) = (D_1/D_3)^2 = 1.44; \quad V_3^2 = 2.07V_1^2 \qquad (8.35)$$

Substitution in Equation (8.26):

$$[(300 - 101)(1000)]/[(1000)(9.81)] = (V_1^2/2g)[f_1(833) + (0.24)(5.06) + f_2(1000)(5.06)$$
$$+ (0.13)(5.06) + f_3(880)(2.07) + 2.07] \qquad (8.36)$$

The values of friction factor at high Reynolds are constant. Referring to the friction–factor plot, Figure A9-1 in Appendix A9:

$$f_1 = 0.0435; f_2 = 0.052; f_3 = 0.048$$

Thus, Equation (8.36) becomes:

$$20.29 = (V_1^2/2g)[36.24 + 1.21 + 263.12 + 0.15 + 87.44 + 2.07] \tag{8.37}$$

$$\therefore \quad V_1 = 1.01 \text{ m/s}$$

$$Q = [(\pi/4)D_1^2](V_1) = 0.071 \text{ m}^3/\text{s} \tag{8.38}$$

**8.4.2** The three pipes of Problem 8.4.1 are connected in parallel. If the total discharge is 0.75 m³/s, what are the flow rates in the individual pipes? The pressure drops through each pipe junction are identical.

### Solution

We assume initially a discharge through one of the pipes say, pipe 1.
Let

$$Q_1 = 0.2 \text{ m}^3/\text{s}$$

Then,

$$V_1 = (Q_1)/[(\pi/4)D_1^2] = 2.83 \text{ m/s} \tag{8.39}$$

Reynolds number for pipe 1, $Re_1 = (V_1 D_1)/v = (2.83)(0.30)/10^{-6} = 8.49 \times 10^5 \tag{8.40}$

From Appendix A9, Figure A9-1: $f_1 = 0.0435$.

$$h_{f1} = f_1(L_1/D_1)(V_1^2/2g) = (0.0435)(200/0.20)(2.83)^2/2g = 17.75 \text{ m} \tag{8.41}$$

For pipes 2 and 3, the head causing the flow should be 17.75 m if $V_1$ was correctly chosen.

$$\therefore \quad 17.75 = f_2(L_2/D_2)(V_2^2/2g) \tag{8.42}$$

Assuming $f_2 = 0.052$ (as before) and solving Equation (8.42) for $V_2$: $V_2 = 2.59$ m/s

$$Re_2 = (V_2 D_2)/v = (2.59)(0.20)/10^{-6} = 5.18 \times 10^5$$

From Appendix A9, Figure A9-1: $f_2 = 0.052$, so that the value of $V_2$ remains unchanged and $Q = 0.0814 \text{ m}^3/\text{s}$.

Similarly, for pipe 3:

$$17.75 = f_3(L_3/D_3)(V_3^2/2g) \qquad (8.43)$$

Assuming $f_3 = 0.048$ (as before) and solving Equation (8.43) for $V_3$: $V_3 = 2.87$ m/s

$$Re_3 = (V_3D_3)/\nu = (2.87)(0.25)/10^{-6} = 7.18 \times 10^5 \qquad (8.44)$$

From Appendix A9, Figure A9-1: $f_3 = 0.048$, so that the value of $V_3$ remains unchanged and $Q = 0.1408\,\text{m}^3/\text{s}$.

$$\Sigma Q = 0.2 + 0.0814 + 0.1408 = 0.4222\,\text{m}^3/\text{s} \qquad (8.45)$$

The Qs may now be adjusted proportionately to make the total discharge $= 0.75\,\text{m}^3/\text{s}$.
   That is,

$$Q_1 = (0.2/0.4222)(0.75) = 0.355\,\text{m}^3/\text{s} \qquad (8.46)$$

$$Q_2 = (0.0814/0.4222)(0.75) = 0.145\,\text{m}^3/\text{s} \qquad (8.47)$$

$$Q_3 = (0.1408/0.4222)(0.75) = 0.250\,\text{m}^3/\text{s} \qquad (8.48)$$

The velocities, Reynolds numbers, friction factors, and head losses are now recalculated based on the new figures.
   The new values are: $V_1 = 5.02$ m/s; $V_2 = 4.62$ m/s; $V_3 = 5.09$ m/s. The corresponding Reynolds numbers increase proportionately to the velocities, so that the friction factors remain unchanged because the values remain constant at high Reynolds numbers. The values of $h_f$ for each pipe are:

$$h_{f1} = f_1(L_1/D_1)(V_1^2/2g) = (0.0435)(250/0.30)(5.02)^2/2g = 55.9\,\text{m} \qquad (8.49)$$

$$h_{f2} = f_2(L_2/D_2)(V_2^2/2g) = (0.0520)(200/0.20)(4.62)^2/2g = 56.6\,\text{m} \qquad (8.50)$$

$$h_{f3} = f_3(L_3/D_3)(V_3^2/2g) = (0.0480)(220/0.25)(5.09)^2/2g = 55.8\,\text{m} \qquad (8.51)$$

The heads are approximately equal, so that the volumetric flow rates through each pipe are:

$$Q_1 = 0.355\,\text{m}^3/\text{s}; \quad Q_2 = 0.145\,\text{m}^3/\text{s}; \quad Q_3 = 0.250\,\text{m}^3/\text{s}$$

## Comment

It is fortuitous that the total volumetric flow rate was chosen as $0.75\,\text{m}^3/\text{s}$. If a much lower flow rate were used together with pipes of lower relative roughness, then the friction factors would vary considerably for each set of iterations. This means that many more calculations would have to be done.

**8.4.3** Water flows through the orifice meter as shown in Figure 8-2: $D_1 = 10$ cm, $Q = 0.01$ m$^3$/s. A differential transducer with tappings at points 1 and 2 has a maximum range of 15 kPa (g). What is the diameter of the orifice at maximum pressure drop? Referring to Figure 8-2, what is the head loss between points 1 and 3? The contraction coefficient, $C_C$ of the orifice may be assumed to be 0.62.

## Solution

There are three diameters of interest in this problem: the diameter of the pipe, $D_1$; the diameter of the orifice, $D_O$; and the diameter at position 2. Position 2 is the minimum area position with a jet diameter, $D_2$. Application of the Bernoulli equation between the orifice and position 2 yields a theoretical flow rate, $Q_{TH}$:

$$Q_{TH} = C_C A_O \{[2(p_1 - p_2)g]/[\rho g(1 - (D_2/D_1)^4)]\}^{1/2} \qquad (8.52)$$

The discharge coefficient for the orifice is defined as:

$$C_D = (Q_{AC}/Q_{TH}) \qquad (8.53)$$

where:

$Q_{AC}$ = actual flow rate

So that Equation (8.52) becomes:

$$Q_{AC} = C_D C_C A_O \{[2(p_1 - p_2)g]/[\rho g(1 - (D_2/D_1)^4)]\}^{1/2} \qquad (8.54)$$

The orifice coefficient is defined as: $C_O = C_D C_C$. Thus, Equation (8.54) becomes:

$$Q_{AC} = C_O A_O \{[2(p_1 - p_2)g]/[\rho g(1 - (D_2/D_1)^4)]\}^{1/2} \qquad (8.55)$$

The area of significance is the area of the orifice, so that Equation (8.55) may be written as:

$$Q_{AC} = C_O A_O \{[2(p_1 - p_2)g]/[\rho g(1 - (D_O/D_1)^4)]\}^{1/2} \qquad (8.56)$$

The discrepancy between Equation (8.55) and Equation (8.56) is incorporated into the orifice coefficient. Also, $\beta^2$ is defined as:

$$\beta^2 = (D_O/D_1)^2 = A_O/A_1 \qquad (8.57)$$

The orifice coefficient, $C_O$ may be obtained from Figure A8-7. The Reynolds number of the flow must first be calculated to use this chart.

$$Re = \rho V_{AV} D_1/\mu = 4Q/\pi \nu D_1 \qquad (8.58)$$

where:

$$v = 10^{-6} \, \text{m}^2/\text{s}$$

Substituting values in Equation (8.58), we obtain: $\text{Re} = 1.27 \times 10^5$. The value of $\beta$ may be obtained iteratively or by trial and error. If we assume initially a value of $\beta$ of 0.55, then the value of $C_O$ from Figure A8-6 at $\text{Re} = 1.27 \times 10^5$ is approximately 0.635. Substituting values in Equation (8.56), we obtain the value of $Q_{AC} = 0.0087 \, \text{m}^3/\text{s}$. A further trial with $\beta = 0.60$ yields $Q_{AC} = 0.0108 \, \text{m}^3/\text{s}$. This is close enough, and no further iteration is warranted. Substitution back into Equation (8.48) and solving for $D_O$ yields: $D_O = 0.06 \, \text{m}$ or 6 cm.

The head loss between points 1 and 3 is given by the relation:

$$(p_1 - p_3)/\gamma = (p_1 - p_2)/\gamma - (p_3 - p_2)/\gamma \tag{8.59}$$

The momentum equation may be applied between points 2 and 3, assuming that the flow is steady and there is no pipe friction:

$$(p_3 - p_2) = (\rho Q/A_1)(V_{AV2} - V_{AV3}) \tag{8.60}$$

where:

$V_{AV3} = Q/A_1$  and  $V_{AV2} = Q/A_2 = Q/C_C\beta^2 A_1$
$C_C =$ contraction coefficient of the orifice plate

Equation (8.60) may be written as:

$$(p_3 - p_2) = (\rho Q^2/A_1^2)(1/C_C\beta^2 - 1) \tag{8.61}$$

Substituting numerical values: $(p_3 - p_2) = 5642 \, \text{Pa}$

The pressure drop between points 1 and 3 is therefore $15,000 - 5642 = 9538 \, \text{Pa} = 9.5 \, \text{kPa}$. This corresponds to 0.97 m water.

## Comment

Variation of pressure through the orifice is shown in Figure 8-3. The pressure energy lost because of the orifice meter is the difference in pressure between the two chained lines, that is, corresponding to the difference between the input and output pressure gradients at positions 1 and 3.

## 8.4.4 Flow through a Venturi Meter

A Venturi meter is connected to measure flow on a pump test as shown in Figure 8-9. The throat diameter of the meter is 12 cm, and the pipe ID is 20 cm. The pressure at the throat of the meter is limited because of cavitation to 3 kPa(a). The meter is connected directly upstream of the pump in a pipe 60 m long and 20 cm diameter. At the inlet to the pipe, a constant head of 10 m is maintained and f = 0.015.

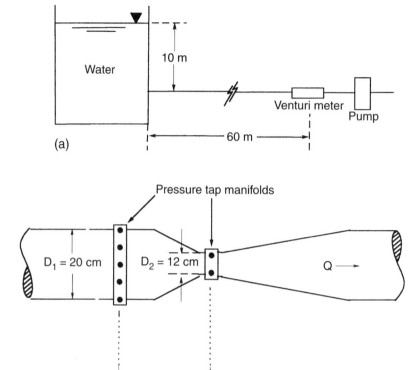

**Figure 8-9** Flow through a Venturi meter. (a) Venturi meter showing position in the system; (b) Detail of Venturi meter.

Calculate:

1. The pressure drop across the meter between points 1 and 2.
2. The maximum discharge into the pump it is permissible for the meter to measure.

Assume that the friction factor for the pipe is constant and equal to 0.0015 and that the discharge coefficient of the meter is equal to 0.96.

**Solution**

Denoting the head loss between 1 and 2 as $h_M$ (meter loss) and applying the steady-state Bernoulli equation between points 1 and 2:

$$p_1/\gamma + V_1^2/2g = p_2/\gamma + V_2^2/2g + h_M \qquad (8.62)$$

Equation (8.62) may be rearranged as:

$$(p_1 - p_2)/\gamma = h = V_1^2/2g[(V_1^2/V_2^2) - 1] + h_M \qquad (8.63)$$

The continuity equation may be written:

$$a_1 V_1 = a_2 V_2 \qquad (8.64)$$

where:

$a_1$ and $a_2$ = cross-sectional areas at 1 and 2.

Combining Equations (8.63) and (8.64) gives:

$$V_1^2 = 2g(h - h_M)/(a_1^2/a_2^2 - 1) \qquad (8.65)$$

The discharge coefficient of the meter:

$$C_D = V_1/V_{1(IDEAL)} = Q_1/Q_{1(IDEAL)} \qquad (8.66)$$

Combining Equations (8.65) and (8.66) yields:

$$h_M = (1 - C_D^2)h \qquad (8.67)$$

The frictional loss along the pipe is given by:

$$h_f = f(L/D)V_1^2/2g \qquad (8.68)$$

Applying the Bernoulli equation between the tank and point 1 and working in absolute pressure values:

$$p_0/\gamma + V_0^2/2g + z_0 = p_1/\gamma + V_1^2/2g + z_1 + f(L/D)V_1^2/2g \qquad (8.69)$$

Substituting values:

$$(101{,}300/9810) + (0) + 10 = p_1/\gamma + V_1^2/2g + (0) + (0.015)(60/0.20)V_1^2/2g \qquad (8.70)$$

$$p_1/\gamma = 20.33 - 5.5V_1^2/2g \qquad (8.71)$$

$$p_2/\gamma = 0.31 \text{ m } [\equiv 3 \text{ kPa(a)}]$$

$$\therefore \quad (p_1 - p_2)/\gamma = h = 20.33 - 5.5(V_1^2/2g) - 0.31 = 20.02 - 5.5(V_1^2/2g) \qquad (8.72)$$

Combining with Equation (8.67):

$$h_M = (1 - 0.96^2)[20.02 - 5.5(V_1^2/2g)] \tag{8.73}$$

$$(h - h_M) = (0.96^2)[20.02 - 5.5(V_1^2/2g)] = (V_1^2/2g)(a_1^2/a_2^2 - 1) \tag{8.74}$$

Noting that $(a_1^2/a_2^2) = (0.2^4/0.12^4) = 7.716$ and substituting in Equation (8.74):

$$V_1 = 5.54 \, \text{m/s and } Q_{MAX} = (\pi/4)(0.2^2)(5.54) = 0.17 \, \text{m}^3/\text{s}$$

**8.4.5** Water flows through the piping system supply as shown in Figure 8-10. The flow out of the system is 0.01 m³/s. The tank is open to atmosphere. The dimensions of the components in the system are: pipe (2–3) 15 cm ID 2 m long; pipe (12–13) 10 cm ID 2 m long; pipe (14–15) 10 cm ID 10 m long; coupling (5–6) 8 cm ID 12 cm long; coupling (7–8) 6 cm ID 10 cm long; basket strainer is 40 cm long. Component (9–10) is a swing-type check valve. The frictional effect of the pipe between the suction surface and point 3 is negligible. The friction factor for all pipes may be taken to be 0.0232. Other data may be found in Appendix A8. A constant head of 10 m from the centerline of the pump to the water level in the tank is maintained. What are the heads at inlet and outlet of the pump, and what power does the pump require at both volumetric flow rates if its overall efficiency is 75%?

**Solution**
The Bernoulli equation is applied between 1 and the liquid level into the tank. Two volumetric flow rates will be considered, and Equation (8.26) is employed:

$$p_1/\gamma + V_1^2/2g + z_1 = p_2/\gamma + V_2^2/2g + z_2 + \Sigma F(\Delta l/D)(V_{AV}^2/2g) + \Sigma K(V_{AV}^2/2g) \tag{8.75}$$

$$\text{Case 1: } Q_1 = 0.1 \, \text{m}^3/\text{s}$$

$$\text{Area pipe (2–3)} = (\pi/4)(0.15^2) = 0.0177 \, \text{m}^2$$

$$\text{Areas of pipes (12–13) and (14–15)} = (\pi/4)(0.10^2) = 0.0079 \, \text{m}^2$$

Equation (8.75) is used in three parts:

1. From the surface of the liquid at suction, position (1) to the inlet of the pump, position (2).
2. Position (2) to pump outlet position (3).
3. Position (3) to the surface of the liquid in the tank, position (4).

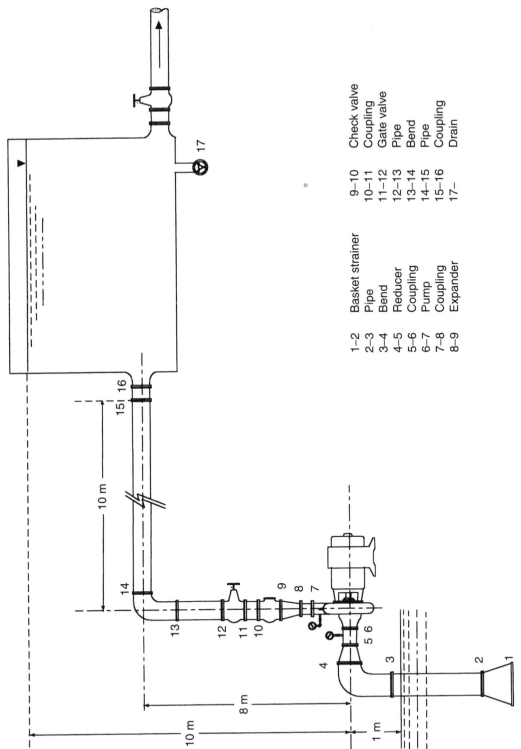

| | |
|---|---|
| 1–2 | Basket strainer |
| 2–3 | Pipe |
| 3–4 | Bend |
| 4–5 | Reducer |
| 5–6 | Coupling |
| 6–7 | Pump |
| 7–8 | Coupling |
| 8–9 | Expander |

| | |
|---|---|
| 9–10 | Check valve |
| 10–11 | Coupling |
| 11–12 | Gate valve |
| 12–13 | Pipe |
| 13–14 | Bend |
| 14–15 | Pipe |
| 15–16 | Coupling |
| 17– | Drain |

**Figure 8-10** Water supply system.

Thus:

$$p_1/\gamma + V_1^2/2g + z_1 = p_2/\gamma + V_2^2/2g + z_2 + \Sigma F(\Delta l/D)(V_{AV}^2/2g) + \Sigma K(V_{AV}^2/2g) \quad (8.76)$$

$$p_2/\gamma + V_2^2/2g + z_2 = p_3/\gamma + V_3^2/2g + z_3 + (1/m)(dW/dt)(1/g) \quad (8.77)$$

$$p_3/\gamma + V_3^2/2g + z_3 = p_4/\gamma + V_4^2/2g + z_4 + \Sigma F(\Delta l/D)(V_{AV}^2/2g) + \Sigma K(V_{AV}^2/2g) \quad (8.78)$$

Substituting values in Equation (8.77):

$$(0) + (0) + (0) = p_2/\gamma + (0.01/0.0177)^2/2g + 1 + (0.0232)(2/0.15)(0.01/0.0177)^2/2g$$
$$+ (1.3 + 0.16 + 0 + 0)(0.01/0.0177)^2/2g + (0.4)(0.01/0.0051)^2/2g \quad (8.79)$$

In Equation (8.79), the reducer and short section loss are considered negligible. Notice that the last term in Equation (8.79) has a different velocity. This is because for a diffuser the velocity used is for the smaller pipe.

$$\therefore \qquad p_2/\gamma = -1.12\,\text{m} \quad (8.80)$$

Combining Equations (8.77) and (8.78):

$$p_2/\gamma + V_2^2/2g + z_2 = p_4/\gamma + V_4^2/2g + z_4 + \Sigma F(\Delta l/D)(V_{AV}^2/2g)$$
$$+ \Sigma K(V_{AV}^2/2g) + (1/m)(dW/dt)(1/g) \quad (8.81)$$

Substituting values in Equation (8.81):

$$-1.12 + (0.01/0.0177)^2/2g + 1 = (0) + (0) + 11 + (0.0232)(2/0.10)(0.01/0.00785)^2/2g$$
$$+ (0.0232)(10/0.10)(0.01/0.00785)^2/2g$$
$$+ (0.4)(0.01/0.0051)^2/2g$$
$$+ (2.5 + 0.015 + 0.16)(0.01/0.00785)^2/2g$$
$$+ 1/(m)(dW/dt)(1/g) \quad (8.82)$$

The mass flow rate, $(dm/dt) = \rho Q = (1000)(0.01) = 10\,\text{kg/s}$.

$$\therefore \qquad -(dW/dt) = 1099\,\text{W or } 1.1\,\text{kW}$$

The negative sign indicates power transferred to the fluid.

Substituting values in Equation (8.77):

$$-1.12 + (0.01/0.0177)^2/2g + 1 = p_3/\gamma + (0.01/0.0051)^2/2g + 1 - (1/10)(1099)(1/g) \tag{8.83}$$

$$p_3/\gamma = 9.90\,\text{m}$$

The actual power required by the pump is $1099/0.75 = 1465$ W or 1.465 kW.

**8.4.6** A large tank is connected in the manner shown in Figure 8-11. The suction head is constant, and in the first case, a well-rounded nozzle (K = 0.04) is attached to the tank. In the second case, a diffuser is attached to the nozzle, (d/D) = 2. $N/R_1 = N/(d/2) = 6$. Compare the flow rates in two cases:

1. When the nozzle is connected directly to the tank.
2. When a diffuser having a diameter ratio of d/D is connected to the nozzle.

## Solution

Referring to the water surface as 0, and the throat of the nozzle as 1, and the end of the diffuser as 2, we may write the Bernoulli equation between the surface water in the tank and the throat of the nozzle:

$$p_0/\gamma + V_0^2/2g + z_0 = p_1/\gamma + V_1^2/2g + z_1 + K_{\text{NOZZLE}}(V_1^2/2g) \tag{8.84}$$

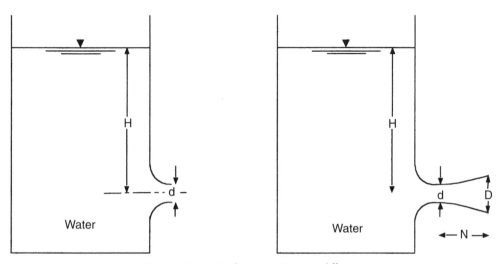

**Figure 8-11** Tank connection to a diffuser.

Substituting values:

$$(0) + (0) + H = (0) + V_1^2/2g + (0) + K_{NOZZLE}(V_1^2/2g) \tag{8.85}$$

From Appendix 8, $K_{NOZZLE} = 0.04$

$$V_1 = [(2gH)/1.04]^{1/2} \tag{8.86}$$

$$Q = [(\pi/4)d^2][(2gH)/1.04]^{1/2} \tag{8.87}$$

Similarly, for the diffuser:

$$p_0/\gamma + V_0^2/2g + z_0 = p_2/\gamma + V_2^2/2g + z_2 + K_{NOZZLE}(V_2^2/2g) + K_{DIFF}(V_2^2/2g) \tag{8.88}$$

Substituting values:

$$(0) + (0) + H = (0) + V_2^2/2g + (0) + 0.04(V_2^2/2g) + K_{DIFF}(V_1^2/2g) \tag{8.89}$$

The value of $K_{DIFF}$ is evaluated from the diffuser chart of Figure A8-8. The area ratio $AR = 4$. From the chart, $C_p = 0.6 = 1 - K_{DIFF}$. Note that the head loss for a diffuser is given in terms of $V_1$ the approach velocity rather than $V_2$. From continuity, $A_1V_1 = A_2V_2$. Therefore, $V_{21} = 4V_2$. Equation (8.89) becomes:

$$H = V_2^2/2g + 0.04(V_2^2/2g) + 0.4((4V_2)^2/2g) \tag{8.90}$$

$$V_2 = \{H(2g)/(2.64)\}^{1/2} \tag{8.91}$$

$$Q = [(\pi/4)D^2]\{H(2g)/(2.64)\}^{1/2} \tag{8.92}$$

$$\therefore \quad Q(nozzle)/Q(diffuser) = (d/D)^2[2.64/1.04]^{1/2} = 1.59(d/D)^2 = 0.4 \tag{8.93}$$

It may be seen that the addition of a diffuser increases the flow rate.

## 8.5 References

Nekrasov, B. *Hydraulics*. Peace Publishers, Moscow (1968), pp. 95–101.

Round, G.F., and Garg, V.K., *Applications of Fluid Dynamics*, Edward Arnold, London (1986).

# CAVITATION

## 9.1 Causes of Cavitation and Parts Affected

When the pressure in a liquid is reduced to the point where the vapor pressure is reached, boiling occurs in the liquid and vapor pockets or bubbles appear. If such bubbles are generated in a turbo machine and carried to a region where the pressure is higher, resulting in collapse of the bubbles, the process is called *cavitation*. Bubbles collapsing on a solid boundary can cause severe damage to the surface (pitting). Photoelastic measurements have shown the pressures generated during this time to be of the order of $1.4 \times 10^9$ Pa. (Sutton, 1957). The lifetime of such bubbles, being of the order of 0.0006 to 0.003 seconds, is also short. Care must therefore be taken to design and operate turbomachines so that cavitation does not occur.

In a turbomachine, reduction in pressure can be brought about in three ways:

1. The liquid level may be lowered by gravity.
2. Frictional dissipation due to shear forces may cause a drop in pressure.
3. The liquid may be accelerated to higher velocities, converting static pressure to kinetic energy.

The nature of the surface is also important in initiating bubble formation. A microscopic bubble has a small radius of curvature. Such bubbles will be trapped in small surface scratches, fissures, or cracks. The pressure in such a bubble must be higher than that in the surrounding liquid by $(2\sigma_S/r)$ where $\sigma_S$ = surface tension of the liquid and r = bubble radius. Bubble formation and entrapment in a fissure are illustrated in Figure 9-1.

In most conditions in which turbomachines are used, there are sufficient surface scratches and nuclei in the liquid that bubbles can form at a little below the vapor pressure of the liquid. However, if pressure reduction occurs only in a very localized region, then nuclei may not be found at this location and further reduction in pressure is possible before cavitation occurs. Bubble growth is also inertia limited because, for a bubble to grow, the liquid surrounding it must acquire an outward radial velocity. In the early-growth stages, the acceleration is small so that bubble collapse may occur at small bubble radii or at a distance from a solid surface. Thus, pressure waves at the surface may be very small.

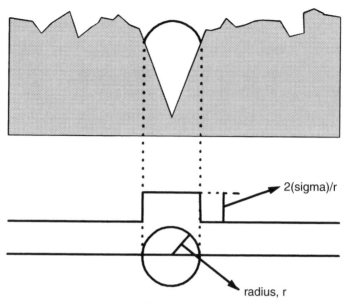

**Figure 9-1** Trapped vapor bubble and pressure induced by surface tension.

Cavitation and erosion are different phenomena, although the effects are the same in terms of metal surface damage. Erosion consists of abrasion of metal walls by particles carried by a liquid, for example, pumping suspensions in a liquid or high-pressure jets of liquid in a machine for example, those generated in high-pressure boiler feed pumps.

### 9.1.1 Methods of Detecting Cavitation

Probably the best methods of detecting cavitation are still by:

1. Direct visual observation of bubble formation
2. Audible detection, unaided or by stethoscope

Other methods are:

1. Change in performance of the machine in terms of head, power, and efficiency
2. Observation and measurement of noise and vibration levels during operation
3. Observation of erosion of parts of the machine after operation for a period of time

## 9.2 Cavitation in Turbines

For turbines, cavitation is expected to be on the low-pressure side of the runner; therefore, suction head is important. However, suction head alone does not determine cavitation. This suction head

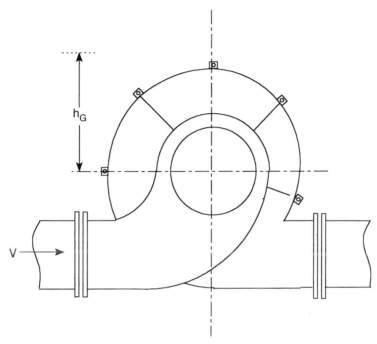

**Figure 9-2** Machine in which the velocity head must be included in $H_{SV}$.

is determined by

$$H_{SV} = h_S + h_A - h_V \qquad (9.1)$$

If there is a static head that is either positive or negative on the runner and this comes from a headwater that is sufficiently large to be able to ignore its velocity, then Equation (9.1) may be used. If, however, the runner is not sufficiently far away from the headwater, then the velocity head must be included in the equation that is,

$$H_{SV} = h_{ST} + h_G + V^2/2g + h_A - h_V \qquad (9.2)$$

The term $(h_{ST} + h_G + V^2/2g)$ is equivalent to $h_S$ in Equation (9.1) and is the net positive suction head NPSH. This definition is valid only when the vertical dimensions of the runner are small compared with $H_{SV}$. Figure 9-2 illustrates the use of Equation (9.2).

In any machine it is necessary to estimate the point at which cavitation may be expected to occur. For example, in a large axial-flow pump or bulb turbine, such a point would be the highest point on the impeller or runner and not on the shaft centerline.

The lower the level that can be maintained at the inlet, the better in terms of cavitation resistance. This is more of a problem with turbines than with pumps because of the added costs of excavation. Because this may be prohibitively high in terms of the overall cost of the system for a turbine installation, a compromise must be reached in terms of minimizing these costs and achieving an acceptable power level and at the same time avoiding cavitation.

## 9.2.1 Thoma Number, $\sigma$

The single most important dimensionless number defining the onset of cavitation is the Thoma number, $\sigma$ (1924/1935). It is defined by the equation:

$$H_S = H_{ATM} - (\sigma H_{NET} + H_{VAP} + H_1) \tag{9.3}$$

where:

$\quad H_{ATM}$ = atmospheric pressure
$\quad H_{NET}$ = net head across turbine
$\quad H_{VAP}$ = vapor pressure of water at prevailing temperature (see Appendix A11)
$\quad\quad H_1$ = height of runner blade above centerline, usually taken to be = $(0.15)(D_1)$
$\quad\quad D_1$ = runner diameter
$\quad\quad\, \sigma$ = cavitation factor or Thoma number

The value of the Thoma number is dependent on several factors:

1. The shape and structure of the turbine
2. The specific speed of the turbine
3. Blade loading

As an illustration of the first two factors, Figures 9-3 and 9-4 show how Thoma number, $\sigma$, varies with head. Figure 9-3 is representative of a large amount of data obtained from working Francis turbines, whereas Figure 9-4 is for axial-flow machines. In all cases, $\sigma$ decreases with increasing head. For geometrically similar machines operating at maximum efficiency, the critical value of $\sigma$ is the same. The specific speed of a machine is related to its shape, so we would expect a correlation between $\sigma$ and $N_S$.

As specific speeds increase, blade loading also tends to increase, for the following reasons:

1. As the requirement for the highest efficiency is aimed for, frictional losses should be kept to a minimum.
2. Minimum blade area is necessary because of 1.
3. There is a practical requirement to keep the head as high as possible to achieve higher flows.

In practical terms, this translates into the critical $\sigma$ increasing with specific speed, which means that the height of high-speed machines above tailwater will be less than that for low-speed machines. In some cases, very high-speed machines must be below the tailwater level.

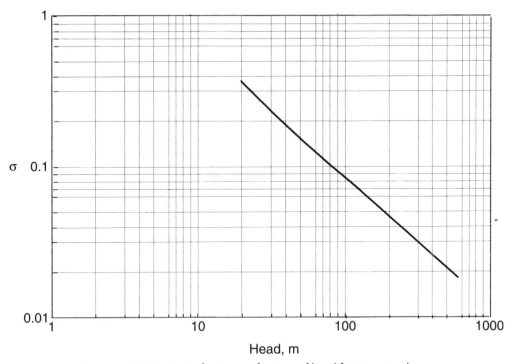

**Figure 9-3** Cavitation factor as a function of head for Francis turbines.

## 9.3 Cavitation in Pumps

Considering, for example, the suction zone of a centrifugal pump, the rotational effect of the blades has a greater impact on the liquid as it approaches the impeller. A tangential velocity component $V_\theta$ to the flow begins and starts to grow. The resultant velocity—the vector addition of $c_\theta$ to the axial component $c_x$ gives an increasing absolute velocity $c_x$, which in turn causes a decrease in the static pressure. Static pressure will continue to fall until the liquid is inside the impeller passage. It is in this region that cavitation may occur; see Figure 9-5. The regions susceptible to cavitation in a centrifugal pump are shown in Figure 9-6.

In a similar way to turbines, Thoma defined a criterion for the onset of cavitation for pumps. It is:

$$\sigma = (p_{ATM}/\rho_g - p_{VAP}/\rho_g) - Z_s)/H \qquad (9.4)$$

H = head across the pump
$Z_s$ = suction head

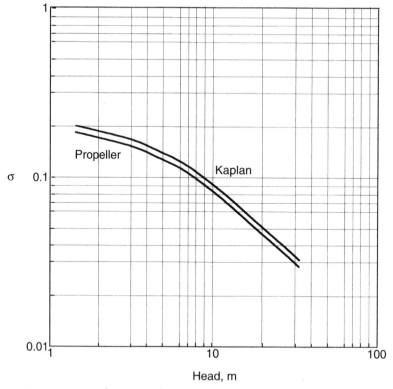

**Figure 9-4** Cavitation factor as a function of head for Kaplan and propeller turbines.

Equation (9.4) is similar to Equation (9.3). It may be seen that the numerator on the right-hand side of this equation is by definition the NPSH. For cavitation, similarity considerations do not take into account the effects of viscosity, surface tension, and compressibility.

### 9.3.1 Cavitation and Specific Speed

A general equation for the mean value of $\sigma$ for impellers as a function of $N_S$ is:

$$\sigma = \text{const. } (N_S^{4/3})\tag{9.5}$$

Wislicenus (1949), on the basis of experimental research, presented two equations: for single-entry impeller pumps,

$$\sigma = 12.2 \times 10^{-4}(N_S^{4/3})\tag{9.6}$$

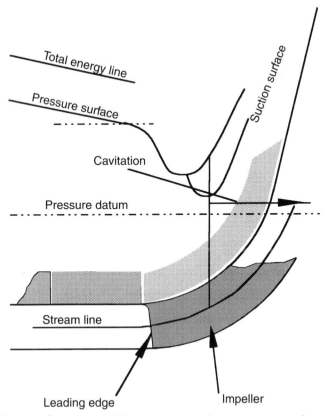

**Figure 9-5** Changes of pressure and kinetic energy in the suction region of a centrifugal pump.

and for double-entry impeller pumps:

$$\sigma = 7.7 \times 10^{-4}(N_S^{4/3}) \tag{9.7}$$

A useful general chart for the region of critical $\sigma$ as a function of specific speed for centrifugal pumps is shown in Figure 9-7.

## 9.4 Determination of Limits of Cavitation

1. Turbines

   The primary method for cavitation detection is to gradually reduce the total suction head under constant operating conditions. Any changes in head, power, and efficiency can be attributed directly to cavitation. The point at which changes are perceptible is called the inception point.

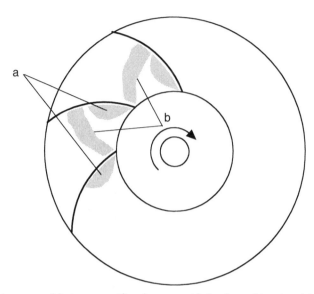

**Figure 9-6** Regions a and b in a centrifugal pump with backward-leaning blades, susceptible to cavitation.

**Figure 9-7** Critical Thoma number, σ as a function of specific speed (SI units).

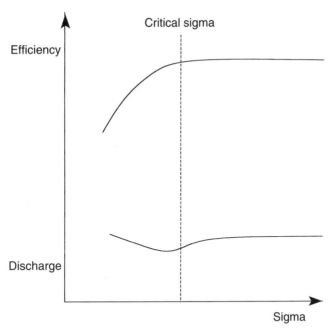

**Figure 9-8** Critical sigma determination for turbines.

2. Pumps

For pumps, the conventional H-Q characteristic curve is used. Incipient cavitation is the upper limit of pump capacity at a given suction head. Figure 9-8 illustrates incipient cavitation for turbines and Figure 9-9 illustrates cavitation effect on centrifugal pumps.

The beginning of cavitation does not necessarily cause a reduction in the fluid dynamic properties of the machine (i.e, head, power, and efficiency). It has been observed for turbines that on occasion, just prior to final collapse of the important properties of the system, the opposite is true. This phenomenon has been known for a long time (Kempf and Foerster, 1932). It has been suggested that this is due to a slight decrease in drag coefficient and a slight increase in lift coefficient on the runner vanes due to vapor formation.

Apart from the Thoma number for characterizing cavitation, the suction specific speed S may be used:

$$S = NQ^{1/2}/H_{SV} \tag{9.8}$$

Like the Thoma number, this may be used for both turbines or pumps. However, Wislicenus (1949) has suggested that this equation may be restricted to the suction passages of the machine.

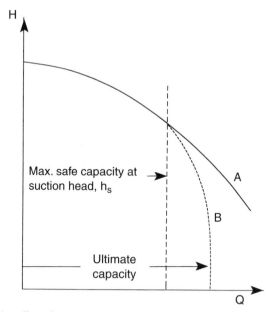

**Figure 9-9** The effect of cavitation on the characteristic curve for centrifugal pumps.

## 9.5 Limitations of Similarity Laws

Similarity equations do not include any viscosity effects or residence time effects of vapor bubbles. In the case of viscosity, any cavitation parameter should include a Reynolds number term. The problem is further complicated for pumps compared with turbines in that pumps may be pumping a wide variety of fluids with very different characteristics, ranging from Newtoniam liquids to time-dependent viscoelastic fluids. So the physical characteristics of the fluid may have a profound effect on cavitation. What pumps and turbines do have in common as parameters are the material of construction and the surface finish.

The two parameters that are usually neglected in considerations of cavitation are:

1. Fluid viscosity
2. Fluid compressibility

Fluids with viscosities that are appreciably higher than those of water must have cavitation phenomena that are Reynolds number dependent. Furthermore, highly viscous fluids do not have a defined single-vapor pressure. The collapse of a vapor bubble is also a function of the residence time of a bubble in a low-pressure zone. The size of a machine will increase the time of bubble residence. The fluid velocities will increase linearly, and if bubble growth is proportional to time, then similarity will be maintained. Because of the surface tension effect, this is not true. Thus, both viscosity and surface tension have an effect on bubble growth.

Another parameter not taken into account is compressibility. Because we are dealing with essentially incompressible fluids, compressibility would appear to be unimportant. While this is true in parts of a pump where bubbles collapse, it is not true in other parts of a pump. Consideration of compressibility leads to the conclusion that for fluids of fixed acoustic velocity the pressures generated by bubble collapse increase according to the first power of fluid velocities. Application of similarity characteristics would indicate that bubble collapse should increase according to the square of fluid velocities.

## 9.6 Methods of Prevention of Cavitation

Increasing the vapor or gas concentration in the liquid lessens cavitation damage. For minimum cavitation damage, the runner rotor material should be resistant to corrosion by the liquid. Other desirable characteristics of the material are:

1. High tensile strength (Ni-Cr stainless steels have been found to be most suitable materials.)
2. High fatigue strength
3. High hardness
4. High resilience

## 9.7 Conclusions about Cavitation

From a number of experimental investigations in this field (Gulich, 1989; Hunsacker, 1935; Poulter, 1942), it may be concluded that for pumps:

1. Cavitation is governed by entrained nuclei (vapor bubbles and particles). There are in every case enough nuclei for bubble formation.
2. The lower the viscosity of the liquid, the easier it is to penetrate the surface pores of the metal; for example, penetration by water is deeper than that of oils.
3. The higher the pressure, the deeper and quicker is the penetration.
4. The smaller the pore area, the greater is the pressure produced when the bubble collapses.
5. The higher the frequency of vibration of metal parts, the more intensive is the destruction.
6. Cavitation is likely to occur even in well-designed pumps when there is flow recirculating at the inlet in low-flow conditions of operation.
7. Cavitation damage increases as the impeller tip speed to the sixth power and the NPSH to the third power.

Research has also shown that for pumps, efficiency increases slightly just before the onset of cavitation.

For both turbines and pumps, intermittent cavitation is not as destructive as strong constant cavitation. The latter type can cause vibration so strong that mechanical failure ensues, and when sufficiently developed complete hydraulic performance breakdown can result.

# 9.8 References

Gulich, J.F., "Guidelines for the prevention of cavitation in centrifugal feed pumps." Final Rept. No. GS-6398 , Sulzer Bros. Ltd. for EPRI, Palo Alto, California (1989).

Hunsacker, J.C., "Cavitation research." *Mech. Eng.* 57 (1935).

Kempf, G., and Foerster, E. (Eds.), "Hydromechanische Probleme des Schiffsantriebs," *Proc. Konferenz über Hydromechanische Probleme des Schiffsantriebs*, 477 pp., Hamburg (1932).

Norrie, D.H., *An Introduction to Incompressible Flow Machines*, Edward Arnold, London (1963).

Poulter, T.C., "The mechanism of cavitation erosion." *Trans. ASME* 9 (1942).

Sutton, G.W., "A photoelastic study of strain waves caused by cavitation." *J. Appl. Mech.* 34, 340 (1957).

Thoma, D., *Bericht zur weltkraftkonferenz*, London,1924. VDI-Zeitschrift, Bd. 79, Berlin (1935).

Wislicenus, G.F., *Fluid mechanics of turbomachinery*. McGraw-Hill, New York (1949).

# WATER HAMMER

## 10.1 Introduction

In this chapter we are concerned with the effects of rapid valve closure in pipes connected to wave reflection points (e.g., reservoirs, pumps, and turbines and rapid starting and stopping of turbomachines). These turbomachines are connected in turn via conduits to wave reflection points. The pressure energy generated by these actions may destroy or severely damage parts of the system. The energy is of two kinds: the kinetic energy of the moving liquid and the elastic energy stored in the liquid and pipes. Both forms are converted to pressure energy, and the rapidity of the conversion is of the utmost importance in terms of the ensuing damage that may result. Such energy must be dissipated in a controlled, nondamaging way.

We consider first the case of instantaneous valve closure at the end of a horizontal pipeline with a flowing liquid, as shown in the sequence of events illustrated in Figure 10-1. The valve is at the right-hand side of the pipeline with a reservoir at the left with a head H, which supplies the necessary potential energy for the flow.

The sequence of events after valve closure is as follows:

1. A wave of positive pressure is generated at the valve and travels upstream (to the left of the valve) with the velocity of sound as in Figure 10-1 (1). During this time, behind the wave the pipe is expanded elastically in the redial direction.
2. When the wave reaches the reservoir, shown in Figure 10-1 (2), the wave pressure falls to the reservoir pressure. The reservoir acts as a reflecting surface.
3. A negative pressure wave now travels downstream with the velocity of sound, and behind the wave the pipe contracts. The fluid velocity behind the wave is negative, and in front of the wave it is zero.
4. When the wave reaches the right-hand side, the valve, it is reflected upstream.
5. The velocity behind the wave traveling toward the reservoir is zero.
6. The negative wave reaches the reservoir, and the pressure rises to reservoir level.
7. A positive reflected wave travels toward the valve.
8. The reflected wave reaches the valve, and one cycle is completed.

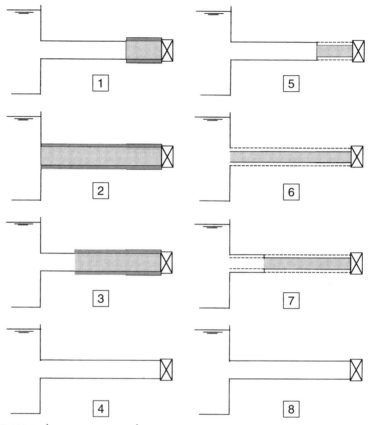

**Figure 10-1** Water hammer-generated wave.
1. Wave propagation upstream immediately after valve closure; 2. wave reaches the reservoir, pipe fully expanded; 3. reflection at reservoir with a negative velocity in the fluid; 4. refection of negative wave at the valve; 5. propagation of negative wave upstream; 6. negative wave at reservoir; 7. negative wave reflected at reservoir; 8. reflected wave reaches the valve, one cycle completed.

Figure 10-2 shows the magnitude of wave velocities in pipes made of steel and cast iron as a function of pipe size.

## 10.2 Equations Describing Wave Generation and Propagation

The equations describing the relationship of pressure, liquid velocity, and wave velocity are:

$$(1/\rho Q)(M^2 V/Mx^2) = (M^2 V/Mt^2) \qquad (10.1)$$

$$(1/\rho Q)(M^2 p/Mx^2) = (M^2 p/Mt^2) \qquad (10.2)$$

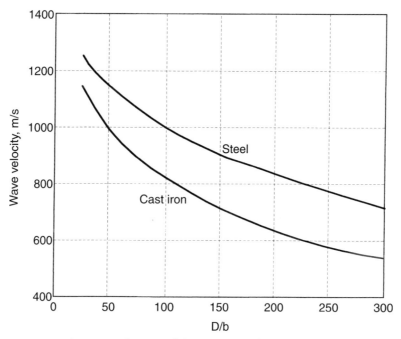

**Figure 10-2** Wave velocity as a function of the ratio (pipe diameter/pipe thickness), D/b, for steel and cast iron pipes.

where:

$\rho$ = liquid density
V = velocity
Q = (1/k + D/bE)
k = bulk compressibility modulus
D = pipe diameter
b = pipe wall thickness
E = modulus of elasticity of pipe wall material

The wave velocity, a, is related to $\rho$ and Q by means of the equation:

$$a = 1/(\rho Q)^{0.5} \qquad (10.3)$$

Equations (10.1) and (10.2) may be manipulated (Rich, 1945) to give:

$$(a^2)(M^2 p/Mx^2) = (M^2 p/Mt^2) \qquad (10.4)$$

$$-(Mp/Mx) = \rho(MV/Mt) \qquad (10.5)$$

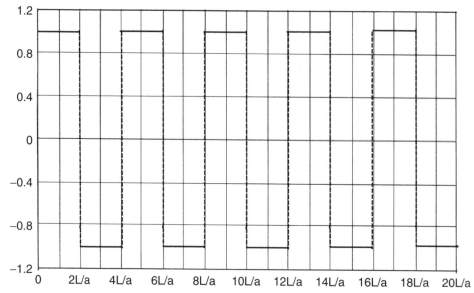

**Figure 10-3** Pressure of wave in conduit as a function of time. The time intervals are given as one-cycle intervals. The maximum and minimum values of pressure = $+\rho a V_0$ and $-\rho a V_0$.

The boundary conditions are: at $x = 0$; $p = p_0$ and at $x = L$; $V = 0$.

Equations (10.4) and (10.5) apply to frictionless flow. Rich (1945) has presented solutions of these equations in terms of series summations of two wave components. Plots of pressure and velocity as a function of time solutions of Equations (10.4) and (10.5) are presented in Figures 10-3 and 10-4. The time intervals are given as one-cycle intervals—that is, the time taken for the wave to travel from the valve to the reservoir and back to the valve. Since friction has not been included in these solutions, the waves are not attenuated.

Figures 10-5 and 10-6 indicate the effects of friction on the attenuation of the waves. In this case, the friction factor has been greatly exaggerated to show the effect. For all practical purposes, in most analyses the magnitude of friction would be such that it would have no appreciable effect on the water hammer pressure and velocities and might be safely ignored. The plot is similar to Figure 10-4 except that the effects of friction are included.

## 10.2.1 Valve Opening or Closure Position as a Function of Time

The position of the end valve in terms of its effective area of opening has a marked effect on wave reflection. The effective area of opening also depends on the type of valve. Figure 10-7 shows the effective area change in terms of valve type and degree of opening.

Types C and D are very close in terms of their characteristics, and very little error would be incurred by assuming the same curve for each valve.

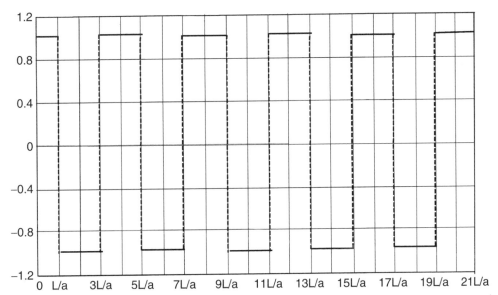

**Figure 10-4** Velocity in conduit as a function of time. The time intervals are given as one-cycle intervals, beginning at L/a. The velocity = velocity at reservoir, x = 0.

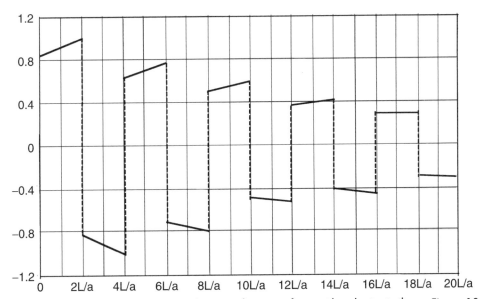

**Figure 10-5** Pressure of wave in conduit as a function of time. The plot is similar to Figure 10-3 except that the effects of friction are included.

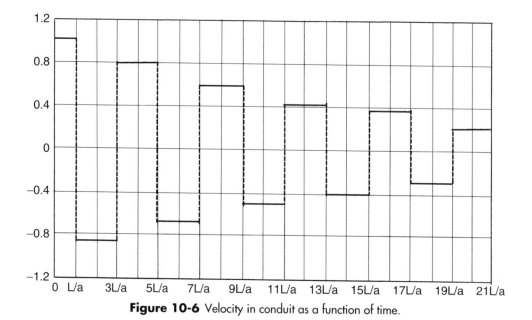

**Figure 10-6** Velocity in conduit as a function of time.

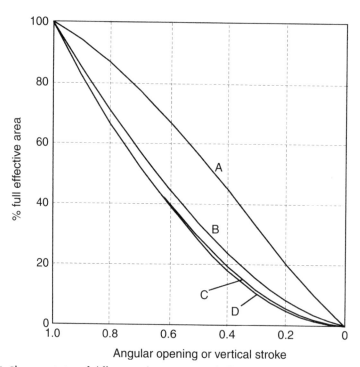

**Figure 10-7** Characteristics of different valves in terms of effective area of flow and closure position.
A—Disk gate valve; B—ring follower gate valve; C—plug valve; D—butterfly valve

# 10.3 Graphical Solution

Undoubtedly the pioneer in the theory of water hammer is Allievi (1925) who developed charts for the series equations, describing the flow. Unfortunately the method is somewhat cumbersome to use. Angus (1935) developed a more rapid graphical method. The uses of Allievi charts and other graphical methods have since been superseded by computer solutions, but because the graphical method is a useful illustration of the solution of the differential equations, it will be described in some detail here.

The basic equations are rewritten in the form of head rather than pressure. The convention used by Angus is that $x = 0$ at the valve and $x = L$ at the reservoir. The basic equations have the form:

$$-(\partial H/\partial x) = (1/g)(\partial V/\partial t) \tag{10.6}$$

$$-(\partial V/\partial x) = (g/a^2)(\partial H/\partial t) \tag{10.7}$$

The general solution of Equations (10.6) and (10.7) is:

$$H - H_0 = F(t - x/a) + f(t + x/a) \tag{10.8}$$

and

$$V_0 - V = (g/a)F(t - x/a) - f(t + x/a) \tag{10.9}$$

The high value of wave velocity compared to cycle time means that intervals of valve gate closure intervals may be executed at cycle time intervals of $(2L/a)$, $(4L/a)$, $(6L/a)$, $(8L/a)$, and so on. Therefore, for successive intervals, Equations (10.8) and (10.9) may be written as:

$$H_1 = H_0 + F_1: H_2 = H_0 + F_2 - F_1: H_3 = H_0 + F_3 - F_2: \ldots \tag{10.10}$$

$$V_1 = V_0 - (g/a)F_1: V_2 = V_0 - (g/a)(F_1 + F_2): V_3 = V_0 - (g/a)(F_2 + F_3): \ldots \tag{10.11}$$

F is a function of the gate setting of the valve; for practical purposes, it may be assumed to be linearly variable. Eliminating F from Equations (10.10) and (10.11), we obtain:

$$H_1 - H_0 = (a/g)(V_0 - V_1) \tag{10.12}$$

$$H_1 - H_2 - 2H_0 = (a/g)(V_1 - V_2) \tag{10.13}$$

$$H_2 - H_3 - 2H_0 = (a/g)(V_2 - V_3) \tag{10.14}$$

$$H_n - H_{n-1} - 2H_0 = (a/g)(V_{n-1} - V_n) \tag{10.15}$$

Equations (10.12) through (10.15) give the relationship between pressure and velocity at the start of each successive interval (2L/a). The function F is the sum of all the positive pressures at time t at position x. Similarly, the function f is the sum of all the negative pressures at time t at position x.

Successively adding and subtracting Equations (10.8) and (10.9):

$$H - H_0 = (-a/g)(V_0 - V) + 2F(t - x/a) \tag{10.16}$$

and

$$H - H_0 = (+a/g)(V_0 - V) + 2f(t + x/a) \tag{10.17}$$

Figure 10-8 presents the notation to be used for direct and reflected wave transmission.

Considering two sections of the pipe A and B, Equation (10.16) may be written for these sections as

$$H_{Bt} - H_{B0} = (-a/g)(V_{B0} - V_{Bt}) + 2F(t - x/a) \tag{10.18}$$

$$H_{At1} - H_{A0} = (-a/g)(V_{A0} - V_{At1}) + 2F(t_1 - x_1/a) \tag{10.19}$$

Since $(t - t_1) = (x - x_1)/a$

$$\therefore \quad F(t - x/a) = F(t_1 - x_1/a) \tag{10.20}$$

The pipe is of uniform diameter; therefore, $V_{A0} = V_{B0}$. We assume $H_{A0} = H_{B0}$. Subtracting Equation (10.19) from Equation (10.18) gives:

$$H_{Bt} - H_{At1} = (+a/g)(V_{Bt} - V_{At1}) \tag{10.21}$$

The same reasoning for the reflected wave yields:

$$H_{Bt1} - H_{B0} = (+a/g)(V_{B0} - V_{Bt1}) + 2f(t_1 + x_1/a) \tag{10.22}$$

$$H_{At} - H_{A0} = (+a/g)(V_{A0} - V_{At}) + 2f(t + x/a) \tag{10.23}$$

Subtracting Equation (10.21) from Equation (10.22) gives:

$$H_{At} - H_{Bt1} = (-a/g)(V_{At} - V_{Bt1}) \tag{10.24}$$

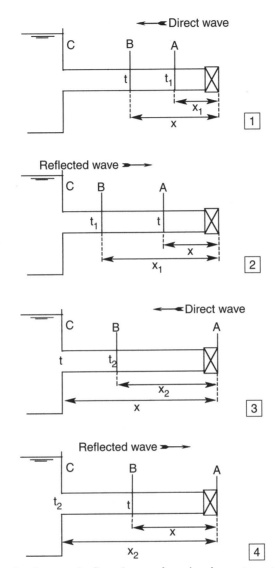

**Figure 10-8** Notation for direct and reflected waves for valve closure in a pipe directly connected to a reservoir.

Equation (10.21) divided by $H_0$ and multiplying the right-hand side by $V_0/V_0$ yields:

$$(H_{Bt}/H_0) - (H_{At1}/H_0) = (aV_0/gH_0)(V_{Bt}/V_0 - V_{At1}/V_0) \tag{10.25}$$

Designating $(aV_0/gH_0)$ by $\rho$, we obtain:

$$h_{At1} - h_{Bt} = +2\rho(v_{At1} - v_{Bt}) \tag{10.26}$$

$$h_{Bt1} - h_{At} = +2\rho(v_{Bt1} - v_{At}) \tag{10.27}$$

$$h_{Ct2} - h_{Bt} = +2\rho(v_{Ct2} - v_{Bt}) \tag{10.28}$$

$$h_{Bt2} - h_{Ct} = +2\rho(v_{Bt2} - v_{Ct}) \tag{10.29}$$

The relation between the gate setting and pipe is:

$$V/V_0 = E(H/H_0)^{1/2} \tag{10.30}$$

where:

E = 1 for a full gate opening.

Equations (10.26) through (10.29) are two series of parallel lines of slope tan $+2\rho$ and tan $-2\rho$. The equations governing the perpetual cycles of after waves are obtained in a similar way. As an example of the variation of head at the valve A as a function of time, let the data for Figure 10-8 be: L = 1524 m, $V_0$ = 3.66 m/s, $H_0$ = 305 m, and wave velocity = 915 m/s. Figure 10-9 is a plot of the variation of head with time at the gate "A."

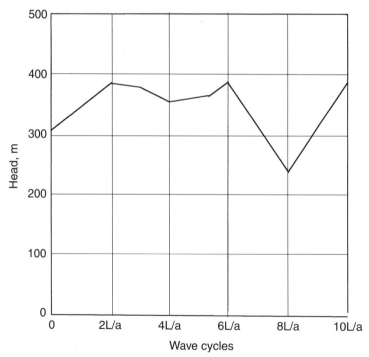

**Figure 10-9** Variation of head in m at the gate of Figure 10-8 as a function of time. Time is shown in wave periods, that is, multiples of 2L/a.

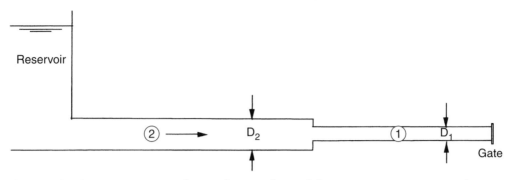

**Figure 10-10** Pressure wave traveling in a horizontal pipe of changing cross section. Pipe 1: diameter = $D_1$: wave velocity = $a_1$; Pipe 2: diameter = $D_2$: wave velocity = $a_2$

## 10.4 Other Wave Reflections

### 10.4.1 Reflection at the Closed End of a Pipe

From Equation (10.9) that is, $V_0 - V = (g/a) F(t - x/a) - f(t + x/a)$

$$V = V_0 = 0 \text{ at any time t.}$$

For a pipe of length L, Equation(10.9) becomes:

$$F(t - L/a) = f(t + L/a) \qquad (10.31)$$

Substitution of Equation (10.31) in Equation (10.8) results in:

$$H - H_0 = 2F(t - L/a) \qquad (10.32)$$

Thus, at the closed end of the pipe the pressure wave is completely reflected without a change of sign, and the magnitude of the reflected wave is twice that of the direct pressure wave.

### 10.4.2 Effect of Change of Area Cross Section

A horizontal pipe with a change of cross-sectional area is shown in Figure 10-10. The velocities of the pressure waves in each pipe are a function of pipe diameter, wall thickness, and pipe material.

Equations (10.8) and (10.9) again are applicable. Considering the junction of the two pipes at point B in Figure 10-10, and labeling the section immediately upstream of B, $B_2$, and the section immediately downstream $B_1$, we may write:

$$H_{B1t} - H_{B10} = F_1 - f_1 \qquad (10.33)$$

$$V_{B1t} - V_{B10} = -(g/a_1)(F_1 - f_1) \qquad (10.34)$$

$$H_{B2t} - H_{B20} = F_2 \qquad (10.35)$$

$$V_{B2t} - V_{B20} = -(g/a_2)F_2 \qquad (10.36)$$

The continuity equation applied to junction B is:

$$A_2 V_{B2t} = A_1 V_{B1t} \qquad (10.37)$$

The velocity head in the two pipe sections may be neglected, so that:

$$F_2 = sF_1 \qquad (10.38)$$

$$f_1 = rF_1 \qquad (10.39)$$

where:

$$s - r = 1 \qquad (10.40)$$

$$s = (2A_1/a_1)/(A_1/a_1 + A_2/a_2) \qquad (10.41)$$

$$r = (A_1/a_1 - A_2/a_2)/(A_1/a_1 + A_2/a_2) \qquad (10.42)$$

$$A_1 = (\pi/4)D_1^2 : A_2 = (\pi/4)D_2^2 \qquad (10.43)$$

s and r are called the transmission and reflection factors, respectively.

### 10.4.3 Junctions and Branches

The derivation of the transmission and reflection factors for pipe branches is similar to that given in Section 10.4.2. Figure 10.11 illustrates a typical branch connection, a bifurcation.

The reflection and transmission factors are given by the following equations:

$$F_2 = F_3 = sF_1 \qquad (10.44)$$

$$f_1 = rF_1 \qquad (10.45)$$

$$s - r = 1 \qquad (10.46)$$

$$s = (2A_1/a_1)/(A_1/a_1 + A_2/a_2 + A_3/a_3) \qquad (10.47)$$

$$r = (A_1/a_1 - A_2/a_2 - A_3/a_3)/(A_1/a_1 + A_2/a_2 + A_3/a_3) \qquad (10.48)$$

$$A_1 = (\pi/4)D_1^2 : A_2 = (\pi/4)D_2^2 : A_3 = (\pi/4)D_3^2 \qquad (10.49)$$

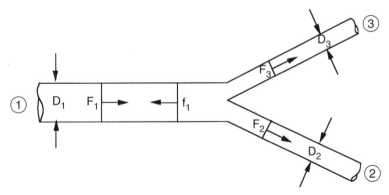

**Figure 10-11** Primary pressure wave and reflected wave traveling along a horizontal pipe that joins two other pipes of different diameter. Pipe 1: diameter = $D_1$: wave velocity = $a_1$; Pipe 2: diameter = $D_2$: wave velocity = $a_2$; Pipe 3: diameter = $D_3$: wave velocity = $a_3$.

The pressure surges transmitted to the branches from the main pipe are equal, irregardless of their cross-sectional areas.

### 10.4.4 Pump Failure

Figures 10-12 to 10-14 illustrate the effects of pump failure. Figure 10-12 shows the sequence of events in terms of wave periods; this is similar in form to Figure 10-1. In this case, a pressure wave is generated at the pump and travels to the reservoir where it is reflected back to the pump. The reflected wave then travels back to the reservoir in an expanded pipe.

## 10.5 Solved Problems

**10.5.1** A liquid of specific weight, $\gamma$, and isothermal bulk modulus, k, flows with an average velocity, V, in a thin-walled pipe of diameter, d, and wall thickness, t. A valve is located at the end of the pipe. The modulus of elasticity, E, and Poisson's ratio, $\sigma$, characterize the pipe material. Derive an expression for the rise in pressure due to rapid closure of the valve assuming the pipe to be elastic.

**Solution**
Let the circumferential stress be denoted by $f_C$ and longitudinal stress by $f_L$.
The volume of fluid in a length x of pipe = $(\pi/4)\, d^2 x$.
    The kinetic energy KE contained in the element of fluid is:

$$KE = \gamma(\pi/4)\, d^2 x (V^2/2g) \qquad (10.50)$$

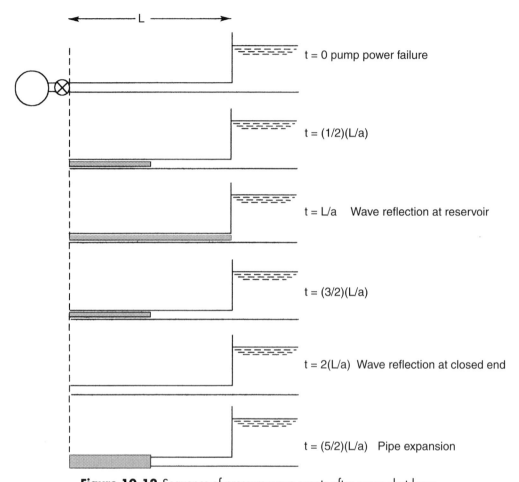

**Figure 10-12** Sequence of pressure wave events after pump shutdown.

The strain energy SE contained in the element of fluid is:

$$SE = \frac{1}{2}\delta p \times \text{volume change} = \frac{1}{2}\delta p \times \Delta V = \frac{1}{2}\delta p(\delta p/k)(\pi/4)\,d^2 x \qquad (10.51)$$

$$\text{Longitudinal strain} = f_L/E - \sigma f_C/E \qquad (10.52)$$

$$\text{Circumferential strain} = f_C/E - \sigma f_L/E \qquad (10.53)$$

Strain energy per unit volume of pipe wall:

$$= \frac{1}{2}f_L(f_L/E - \sigma f_C/E) + \frac{1}{2}f_C(f_C/E - \sigma f_L/E) = (\frac{1}{2}E)(f_L^2 + f_C^2 - 2\sigma f_L f_C) \qquad (10.54)$$

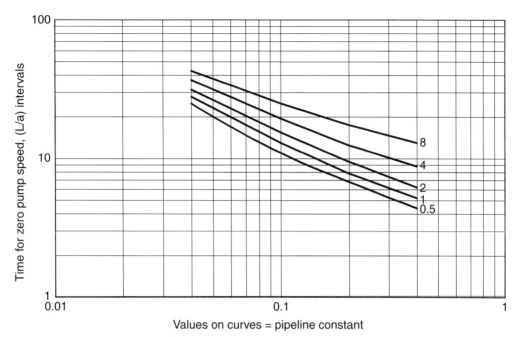

Values on curves = pipeline constant

**Figure 10-13** Plot of time to reach zero pump speed in terms of pipeline constant.

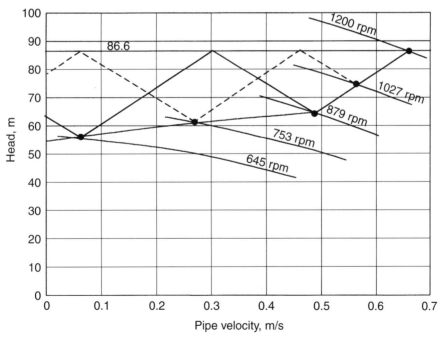

**Figure 10-14** Head versus pipe velocity. Progress of the surge wave can be seen at different pump speeds.

For a thin-walled pipe:

$$f_L = (\delta p \times d)/4t \tag{10.55}$$

$$f_C = (\delta p \times d)/2t \tag{10.56}$$

Strain energy per unit volume of pipe:

$$= (\frac{1}{2}E)\{[(\delta p)^2 d^2)/16t^2] + [(\delta p)^2 d^2)/4t^2] - \sigma[(\delta p)^2 d^2)/4t^2]\} \tag{10.57}$$

Volume of pipe wall of length $x = \pi$ dt x

$$\text{Strain energy of pipe wall of length } x = [\pi d^3 (\delta p)^2 x/(16Et)](5 - 4\sigma) \tag{10.58}$$

$$\text{The original kinetic energy of the fluid } = \text{ strain energy of fluid } + \text{ strain energy of pipe} \tag{10.59}$$

$$\therefore \quad \gamma(\pi/4)d^2 x(V^2/2g) = (\pi/4)d^2 x(\delta p)^2(1/2k) + (\pi/4)d^2 x[(\delta p)^2 d/(4Et)](5 - 4\sigma) \tag{10.60}$$

$$\therefore \quad \delta p = V(\gamma/g)\{1/[(1/k) + d(5 - 4\sigma)/(4tE)]\} \tag{10.61}$$

### 10.5.2 Estimation of Surge Pressure in a Pipe-pump System

A centrifugal pump is attached to a pipe 275 m long with a diameter of 1 m. A nonreturn valve is fitted at the pump. Under normal operating conditions, the pump operates as follows:

N = 1200 rpm; H = 86.6 m; Q = 0.591 $m^3$/s; W = 591 kg/s
Moment of Inertia of the rotating parts = 61.2 kg-$m^2$

When the pump is shut down, estimate the positive and negative surge pressures in the system; what is the pump speed immediately after valve closure? It is assumed that the pump characteristic at 1200 rpm is known and that the pump efficiency at the normal running condition is 85%. Friction in the pipe may be neglected.

### Solution

It is easier and more instructive to make a graphical solution. This is based on the work of Allievi (1925) and Angus (1935) as outlined in Section 10.3.

A commonly used value of the wave velocity in such a system is 1220 m/s. Note that the speed of sound in water under these conditions is 1433 m/s. The time for the wave to travel along the pipe in one direction is therefore $= \Delta t = 275/1220 = 0.225$ s. During this time, the pump is slowing down. The change of rotational speed is given by:

$$dN = [(dt)(K)(W)(H)]/[(\eta_O)(N)(I)] \tag{10.62}$$

**Table 10-1**

| Period | $V_x$ m/s | W kg/s | H, m | $\eta_0$ | N rpm | $\Delta$N rpm |
|--------|-----------|--------|------|----------|-------|---------------|
| 1 | 0.66 | 591 | 86.6 | 0.85 | 1200 | 173 |
| 2 | 0.56 | 506 | 74.1 | " | 1027 | 148 |
| 3 | 0.48 | 433 | 63.4 | " | 879 | 126 |
| 4 | 0.41 | 371 | 54.3 | " | 753 | 108 |
| 5 | 0.35 | 318 | 46.5 | " | 645 | 93 |

where:

$K = 936$
$W$ = mass flow rate
$H$ = head at operating point
$\eta_0$ = overall efficiency of pump
$N$ = rotational speed
$I$ = moment of inertia of rotating parts

The velocity of the fluid in the pipe is:

$$V_P = Q/A_P \tag{10.63}$$

The following relations also apply:

$$V_2 = V_1(N_2/N_1) \tag{10.64}$$

$$W_2 = W_1(N_2/N_1) \tag{10.65}$$

$$H_2 = H_1(N_2/N_1) \tag{10.66}$$

Using Equations (10.62), (10.64), (10.65), and (10.66), the appropriate values for each transit period of the wave may be put in tabular form (see Table 10-1).

The data of Table 10-1 may now be plotted. The resulting waveforms are shown in Figure 10-14. The characteristic curve for the normal operating condition is known, and a section of it is also shown in Figure 10-14. Once the rotational speeds have been established, the other characteristics passing through these points may be drawn. The points on these curves are determined from the values of H and pipe velocity $V_X$. Thus, starting at point 0 on the 1200 rpm curve, the next point is x on the 1027 rpm curve. Point x is located at $H = 71.4$ and $V_x = 0.56$. The construction lines are drawn from the slope given by $(V_0/g)$.

Starting at point 0, this line intersects the 1027 rpm curve at x. The other construction lines are similarly drawn. The points $a_1$, $a_2$, and $a_3$ fall on the other characteristic curves as shown.

The value of the head at a pipe velocity = zero is approximately 55 m. The time for the flow to cease through the pump is about 4.3–4.4 periods, that is, approximately 1 s.

$$\text{Surge pressures are: (1) Negative surge} = 88.6 - 55 = 33.6 \text{ m}$$
$$\text{(2) Positive surge} = 88.6 + 55 = 143.6 \text{ m}$$

## Comment

The fundamental Equation (10.45) is true for only very small speed changes, dN. The greater the value of the time interval used for calculations $\Delta t$, the greater is the discrepancy between conditions at the beginning and middle of the period. However, the result is a conservative estimate in that calculated surge pressures are exaggerated.

# 10.6 References

Allievi, L., *The Theory of Water Hammer*. Garroni, Rome (1925).

Angus, R.W., "Simple graphical solution for pressure rise in pipe and pump discharge lines." *J. Eng. Inst. Canada*, February (1935).

de Haller, P., "Investigation of corrosion phenomena in water turbines." *Escher-Wyss News,* Zurich (1933).

Rich, G.R., "Water-hammer analysis by the Laplace-Mellin Transformation." *Trans. ASME*, Paper 44-A38 (1945).

# CORROSION

## 11.1 Introduction

The principal thing that cavitation and corrosion have in common as far as turbomachines are concerned is that they both cause metal loss. Cavitation is a mechanical mechanism, and corrosion is a chemical one. In addition, corrosion may cause mechanical failure of metal parts, which may have disastrous consequences.

The mechanism of corrosion is an electrochemical reaction of a metal with its environment. With turbines, the environment is aqueous and with pumps, the environment for the most part is aqueous, but a wide variety of fluids may also be pumped. In any case, the possibility of corrosion, severe or otherwise, occurring in a turbomachine or its ancillary equipment cannot be ignored. The consequences of ignoring it may have safety implications and most certainly have economic ones.

In this chapter we concentrate on the corrosion of iron, steel, and its alloys as they relate to use in hydraulic turbines and pumps. A good general reference text on all other corrosion phenomena is the text edited by Uhlig (1948).

## 11.2 Thermodynamics of the Corrosion Process

The thermodynamic explanation of corrosion in its broadest sense is the tendency of a system in a high-energy state to transform, by interaction with its surroundings, to a low-energy state. It is useful at this stage to define a thermodynamic function that will quantify this free energy change.

This function is called the Gibbs Function, G. It is defined by:

$$G = H - TS \qquad (11.1)$$

where:

H = enthalpy
T = temperature
S = entropy

In terms of specific properties:

$$g = h - Ts \tag{11.2}$$

Differentiating Equation (11.2):

$$dg = dh - T\,ds - s\,dT \tag{11.3}$$

and substituting the equation for enthalpy:

$$dh = T\,ds + v\,dp \tag{11.4}$$

we obtain:

$$dg = vdp - sdT \tag{11.5}$$

A finite change in free energy of a corrosion reaction is represented by $\Delta G$. Because G is temperature dependent, values of standard $\Delta G$ at fixed temperature and pressure, 298 K and one atmosphere, are tabulated in standard thermodynamic tables (e.g., International Critical Tables). The standard values are designated as $\Delta G^0$. A discussion of the Gibbs function may be found in any good engineering thermodynamics text, for example, Moran and Shapiro (2000).

The temperature dependence of G is given by the thermodynamic equation:

$$\Delta G = \Delta G^0 + RT \ln J \tag{11.6}$$

where:

$$J = [products] / [reactants]$$

Thus, for the reaction:

$$A + B \rightarrow C + D$$
$$J = [C][D]/[A][B]$$

A chemical reaction at a specified temperature and pressure must always proceed in the direction of decreasing Gibbs function. Equilibrium is reached when:

$$(dG)_{T,P} = 0 \tag{11.7}$$

An increase in Gibbs function during chemical reaction is a contradiction of the Second Law. This is illustrated in Figure 11-1.

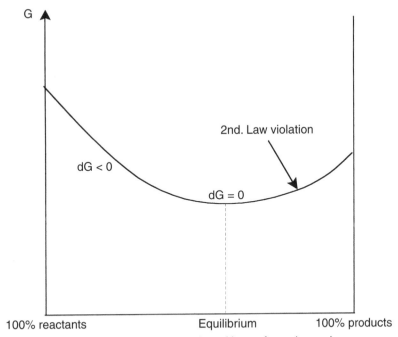

**Figure 11-1** Criterion for chemical equilibrium for a chemical reaction.

A corrosion reaction rate v may be expressed as:

$$v = K \, [reactants] \tag{11.8}$$

where:

$$K = A \exp(-\Delta G/RT)$$

$$A = constant \tag{11.9}$$

The conventional thermodynamic sign system is energy emitted from the system to be denoted negative (−) and energy absorbed by the system as positive (+). For a reaction to occur spontaneously, $\Delta G$ must be negative. An example illustrating the use of Equation (11.6) is the reaction between copper and iron in a solution of copper sulfate. This is a "corrosion cell."

The iron is corroded because it acts as an anode as:

$$Fe \rightarrow Fe^{++} + 2e^- \tag{11.10}$$

The copper, on the other hand, has a cathodic reaction, which is the opposite of the corrosion reaction.

$$Cu^{++} + 2e^- \rightarrow Cu \tag{11.11}$$

The complete reaction is thus:

$$Fe + Cu^{++} \rightarrow Fe^{++} + Cu \tag{11.12}$$

In the case of this system, Equation (11.6) becomes:

$$\Delta G = \Delta G^0 + RT \ln\{[Fe^{++}][Cu]/[Cu^{++}][Fe]\} \tag{11.13}$$

Thus, the free energy is driving the reaction, and the rate of the reaction is governed by the rate of change of free energy.

## 11.3 Corrosion of Iron and Steel

Iron (Fe) as an anode has a reaction according to Equation (11.10). Free hydrogen ions in the electrolyte have a cathodic reaction:

$$H^+ \rightarrow 1/2H_2 - e^- \tag{11.14}$$

This is a rapid reaction in acids but slow in neutral or alkaline media. The cathodic reaction can be accelerated with dissolved oxygen. Thus:

$$2H^+ + 1/2O_2 \rightarrow H_2O - 2e^- \tag{11.15}$$

The reaction of iron with water and dissolved oxygen becomes:

$$Fe + H_2O + 1/2O_2 \rightarrow Fe(OH)_2 \tag{11.16}$$

Ferrous oxide, Fe $(OH)_2$, becomes a diffusion barrier next to the iron surface through which $O_2$ must diffuse for further reaction to take place with the iron. At the outer surface of the film, the ferrous oxide is further converted by means of the reaction:

$$Fe(OH)_2 + 1/2H_2O + 1/4O_2 \rightarrow Fe(OH)_3 \tag{11.17}$$

Hydrous ferric oxide is orange to red-brown in color and is the familiar "rust." The two forms are nonmagnetic, $\alpha$- $Fe_2O_3$ (Hematite), and magnetic $\gamma$-$Fe_2O_3$. Very often a magnetic hydrous ferrous ferrite, $Fe_3O_4Xn\ H_2O$, is formed between hydrous $Fe_2O_3$ and FeO. Rust usually consists of three layers in different states of oxidation.

### 11.3.1 Effect of Temperature

When oxygen diffusion is the controlling mechanism for corrosion, there is a linear increase of corrosion rate, with temperature in the range of 30° C to 80° C (Speller, 1951). Beyond 80° C there is a rapid falling off of the corrosion rate in open systems. The reason for this is the decreased solubility of oxygen in water as temperature is increased. This effect outweighs the effect of temperature alone. Turbines may be regarded as operating as open systems. On the other hand, it was found that for closed systems the linear increase with temperature continued. Pumps may operate as open or closed systems.

### 11.3.2 Effect of pH

In the pH range of 4 to 10, the rate of corrosion is independent of pH. For turbines then, the effect of pH may be ignored. Pumps, however, have a wide variety of fluids that are pumped, including acids.

For fluids with a pH $< 4$, the ferrous oxide film is dissolved, and iron comes in contact with an aqueous environment again. As the pH approaches the range $< 3$, the rate of reaction increases very markedly as a result of hydrogen evolution and oxygen depolarization. The implication of this is that pumps pumping liquids that are markedly acidic should not have any iron parts at all.

### 11.3.3 Action of Anaerobic Bacteria

When iron is immersed in deaerated water, the corrosion rate is relatively small. In some aqueous environments, the corrosion rate is high. This phenomenon was first observed by von Wolzogen Kűhr (1923) and was found to be due to the presence of sulfate-reducing bacteria, *sporovibrio desulfuricans*. The mechanism of this corrosion is by reduction of inorganic sulfates to sulfides in the presence of surface-adsorbed hydrogen provided by the iron.

### 11.3.4 Pitting and Crevice Corrosion

Pitting is a form of local corrosion that starts and is propagated from a surface defect in the metal. The defect may be caused by:

1. A residual stress on the surface during formation of the metal
2. An inclusion on the surface
3. A break or scratch on the surface

Crevice corrosion may be regarded as an extension of pitting. It is a local breakdown of the passive condition. Once a surface has been pitted and a crack or crevice is formed, it is autocatalytic; that is, corrosion is self-sustaining once it begins. Inside the crevice corrodes rapidly, while the outside is cathodically protected. An interesting feature in stainless steel corrosion is that it is not the corrosion of the iron in the steel that is most damaging (Peterson et al. 1970) but the dissolution and hydrolysis, which cause a drop in pH.

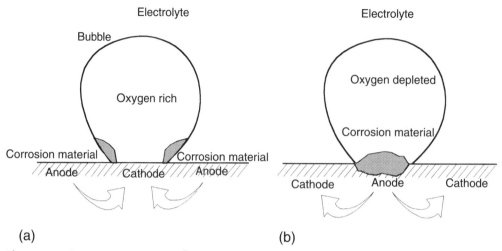

**Figure 11-2** Pitting corrosion mechanism: (a) initiation with oxygen enriched bubble; (b) oxygen-depleted bubble with rust scab formation.

The reaction is:

$$Cr^{+++} + 3H_2O \rightarrow Cr(OH)_3 + 3H^+ \tag{11.18}$$

It has been found that the pH of electrolyte in an active crevice is very low. Crevice corrosion is a local breakdown of the passive condition. In summary, both pitting and crevice corrosion increase with:

1. Increase of medium temperature
2. Lower chromium and molybdenum content of the steel
3. Increase of chloride content of the fluid medium

In the case of chloride content, two mechanisms for increased corrosivity have been proposed. The first is that the chloride ion, $Cl^-$, permeates through defects and pores in the oxide film more easily than other ions such as $SO_4^{--}$. The second suggests that $Cl^-$ ions adsorb on the metal surface more easily than in $O_2$ or $OH^-$. When this is accomplished, metal ions enter into solution more easily. This is opposite to the effect of adsorbed oxygen. The mechanisms of pitting and crevice corrosion are illustrated in Figures 11-2 and 11-3.

## 11.4 Corrosion Resistance of Steel Alloys

Plain carbon steels, that is, nonalloyed steels, corrode easily in most environments. Adverse corrosion conditions include:

1. Plentiful oxygen supply
2. Low pH

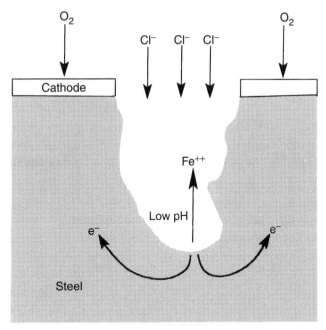

**Figure 11-3** Crevice corrosion mechanism.

3. Presence of ions such as oxides of sulfur and chlorides
4. Increase of water flow rate past the surface

Alloying with chromium, Cr, has the effect of inducing passivity, that is, increasing corrosion resistance. The 8% Ni-austenitic steels are the most popular of all the stainless steels produced.

The three main classes of stainless steels are martensite, ferrite, and austenite.

1. Martensite

    Martensite is produced by rapid quenching of steel from the austenitic region of the phase diagram. The structure of the steel is face-centered cubic, and it is magnetic. Applications include turbine blades and tools. Typically, AISI nos. 403 and 410 are used here.

2. Ferrite

    This stainless steel is named after the ferrite; the $\alpha$-phase for pure iron of carbon steels is cooled slowly from the austenite region.

3. Austenite

    These steels are named after the $\gamma$-phase or austenite, which for pure iron exists at 910–1400° C. Alloyed Ni is largely responsible for retention of austenite when Cr-Fe-Ni alloys are quenched. Mn, Co, C, and N also contribute to the retention and stability of the austenite phase.

The highest corrosion resistance for alloys is obtained with high Ni composition. Mo-containing alloys such as AISI nos. 316 and 316 L have good corrosion resistance to chloride environments and to crevice corrosion. For both turbines and pumps, if a stainless steel is to be used for most components, probably 316 S is to be preferred.

In both ferritic and austenitic stainless steels, the ratio of chromium to molybdenum is very important. For austenitic stainless steels, Cr of 20–22% requires 6% Mo, and for ferritic and stainless steels 25–28% Cr requires 3% Mo.

## 11.5 Stress Corrosion Cracking and Corrosion Fatigue

These types of corrosion are characterized by cracks and brittle failure and occur below the yield point of the original material. To determine the tendency to failure by stress corrosion, crack growth measurements are carried out with and without the environment present (Parkins, 1979). Qualitatively, the data are divided into three regions:

1. A region where crack growth is related to stress intensity but drops very rapidly. In this region, threshold stress intensity occurs, below which there is no crack growth.
2. A region where there is little dependence on stress intensity and crack growth occurs at an almost constant rate.
3. A region where the effect of the environment is negligible and fracture is strongly dependent on stress intensity.

Corrosion fatigue occurs when repeated load cycling acting in conjunction with a corrosive medium causes cracks to appear on the surface of the material. The amount of corrosion is a function of the number of cycles and of the frequency of the load cycle.

## 11.6 Galvanic or Bimetallic Corrosion

Galvanic corrosion occurs when two dissimilar metals are joined together to form a corrosion cell. The standard arrangement of the oxidation or reduction potential is the electrochemical series. Details of the series are given in International Critical Tables.

The more positive oxidation or more negative reduction potentials correspond to more reactive metals. The position of a metal in the series is determined by the equilibrium potential of a metal in contact with its ions at a concentration equal to unit activity. Unfortunately, the series has limited use for practical predictions because unit activity for some metal salts requires concentrations that are impossible to obtain. Hydrogen in this series is assigned the value zero. The more positive emf of a metal means it has a greater oxidation potential. For example, Lithium (Li) has a standard oxidation potential of $+3.05$ volts. This metal is the most reactive of all. At the other end of the scale, Gold (Au) has the largest negative value: $-1.50$ volts. As a result, gold has the lowest potential for corrosion. It is termed a *noble* metal.

Because of the limitations of the emf series, a so-called Galvanic Series has been suggested. This series is arranged for metals and their alloys in accordance with their measured potentials

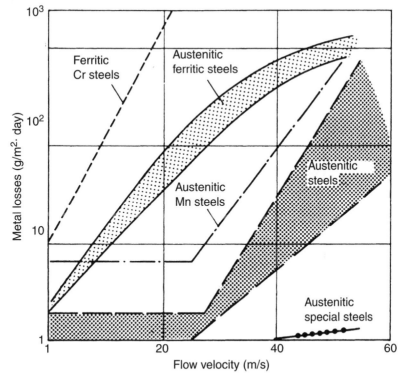

**Figure 11-4** Rate of corrosion of different steels as a function of fluid velocity. (Courtesy Sultzer Pumps Ltd., Zurich)

in a given environment (Uhlig, 1948). There are a number of galvanic series because of different environments. A commonly used galvanic series, for example, has seawater as an environment.

## 11.7 Cathodic Protection

Cathodic protection is the most important technique for corrosion control. The method consists of making sure that no ions flow into the environment—that is, corrosion current flow is reduced to zero. This is done by application of a current to a circuit consisting of the object to be protected, and the cathode connected to an anode through which positive DC current is supplied. The source of current is usually a rectifier supplying low-voltage DC current of several amperes.

### 11.7.1 Sacrificial Anodes

An anode consisting of a metal that is more active in the galvanic series is attached to the object to be protected. Sacrificial anodes are usually magnesium, Mg, or magnesium-base alloys.

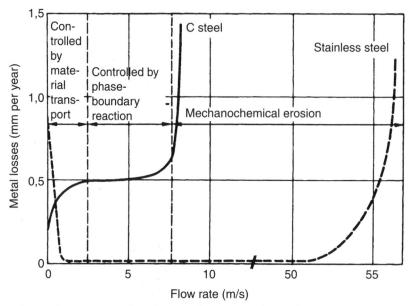

**Figure 11-5** Mechanisms controlling the rate of corrosion of C steel and stainless steel as a function of fluid velocity. (Courtesy Sultzer Pumps Ltd., Zurich)

Other metals such as zinc are used to a lesser extent. If steel is to be protected, the open circuit potential difference for Mg is of the order of 1 volt. The degree of protection is increased by immersion in a high-conductivity environment. The reaction for Mg is:

$$2Mg^+ + 2H_2O \rightarrow Mg(OH)_2 + Mg^{++} + H_2 \qquad (11.19)$$

If the environment has NaCl content, $MgCl_2$ is also formed with $H_2$ evolution at the anode.

### 11.7.2 Protection and Overprotection

One aim of cathodic protection is to make sure that too much current is not used and that the anodes last as long as possible. A moderate amount of overprotection does no harm. A simple test to ascertain the correct amount of current to apply is by the use of test coupons of anodic material to a section of the object to be protected. Weight losses of the coupons are measured to determine the effectiveness of the applied current.

## 11.8 Effect of Flow Rate of the Environmental Fluid

In general, increasing the flow rate of the environmental fluid increases corrosion. There are limiting flow rates for all metals and alloys. Figure 11-4 shows the corrosion rates of different alloy steels as a function of flow velocity. The fluid was water with 230,000 mg/l of dissolved salts, giving a pH of 4.5.

Figure 11-5 shows the corrosion rates of two steels and the controlling mechanisms as a function of flow rate. There is a wide range of flow velocity, up to 50 m/s, with negligible effect on corrosion/erosion.

# 11.9 References

de Haller, P., "Investigation of corrosion phenomena in water turbines." *Escher-Wyss News*, Zurich (1933).

Moran, M.J., and Shapiro, H.N., *Fundamentals of Engineering Thermodynamics* (4th Ed). John Wiley & Sons, New York (2000).

Parkins, R.N., "Development of strain rate testing and its implications," in *Stress Corrosion Cracking: The Slow Strain Rate Technique*. Ed. Ugiansky, G.M., and Payer, J.H., ASTM-STP 665, Philadelphia (1979).

Peterson, M.H., Lennox, T.J., and Groover, R.E., "A study of crevice corrosion in Type 304 stainless steel." *Materials Protection* 9 (no.1), 23–26 (1970).

Poulter, T.C., "The mechanism of cavitation erosion." *Trans. ASME* 9 (1942).

Speller, F., *Corrosion, Causes and Prevention*, McGraw-Hill, New York (1951).

Uhlig, H.H., *Corrosion Handbook*, Ed. H.H.Uhlig, John Wiley & Sons, New York (1948).

von Wolzogen Kűhr, C., *Water and Gas*, 7, No. 26 (1923).

# EQUATIONS

## Continuity Equation

### Cartesian coordinates:

$$\partial\rho/\partial t + \partial(\rho V_x)/\partial x + \partial(\rho V_y)/\partial y + \partial(\rho V_z)/\partial z = 0 \tag{A.1}$$

For a steady-state, inviscid, two-dimensional, incompressible flow:

$$\partial V_x/\partial x + \partial V_y/\partial y = 0 \tag{A.2}$$

### Polar cylindrical coordinates:

$$\partial\rho/\partial t + (1/r)[\partial(\rho r V_r)/\partial r] + (1/r)[\partial(\rho r V_\theta)/\partial\theta] + \partial(\rho V_z)/\partial z = 0 \tag{A.3}$$

For steady-state, inviscid, two-dimensional, incompressible flow:

$$\partial(r V_r)/\partial r + \partial(V_\theta)/\partial\theta = 0 \tag{A.4}$$

## Energy Equation

### Cartesian coordinates:

$$\frac{\partial}{\partial x}\left[k\left(\frac{\partial T}{\partial x}\right)\right] + \frac{\partial}{\partial y}\left[k\left(\frac{\partial T}{\partial y}\right)\right] + \frac{\partial}{\partial z}\left[k\left(\frac{\partial T}{\partial z}\right)\right] + \Phi = \rho\frac{\partial}{\partial t}[C_p T] + \rho V_x\frac{\partial}{\partial x}[C_p T]$$

$$+ \rho V_y\frac{\partial}{\partial y}[C_p T] + \rho V_z\frac{\partial}{\partial z}[C_p T] - \left(\frac{\partial p}{\partial t}\right) + V_x\left(\frac{\partial p}{\partial x}\right) + V_y\left(\frac{\partial p}{\partial y}\right) + V_z\left(\frac{\partial p}{\partial z}\right)$$

$$\tag{A.5}$$

where:

$$\Phi = 2\mu[(\partial V_x/\partial x)^2 + (\partial V_y/\partial y)^2 + (\partial V_z/\partial z)^2 + \frac{1}{2}[(\partial V_x/\partial y) + (\partial V_y/\partial x)]^2$$

$$+ \frac{1}{2}[(\partial V_y/\partial z) + (\partial V_z/\partial y)]^2 + \frac{1}{2}[(\partial V_z/\partial x) + (\partial V_x/\partial z)]^2$$

and

$$h = [C_p T]$$

## Polar cylindrical coordinates:

For a viscous, incompressible flow:

$$(1/r)\frac{\partial}{\partial r}[kr(\partial T/\partial r)] + (1/r^2)\frac{\partial}{\partial \theta}[k(\partial T/\partial \theta)] + \frac{\partial}{\partial z}[k(\partial T/\partial z)] + \Phi = \rho(Dh/Dt - Dp/Dt)$$

$$(A.6)$$

where:

$$\Phi = 2\mu\{(\partial V_t/\partial r)^2 + [(1/r)(\partial V_\theta/\partial y) + (V_r/r)]^2 + (\partial V_z/\partial z)^2$$

$$+ \frac{1}{2}[(\partial V_\theta/\partial r) - (V_\theta/r) + (1/r)(\partial V_t/\partial \theta)]^2$$

$$+ \frac{1}{2}[(1/r)(\partial V_z/\partial \theta) + (\partial V_\theta/\partial z)]^2$$

$$+ \frac{1}{2}[(\partial V_t/\partial z) + (\partial V_x/\partial r)]^2 - \frac{1}{3}(\nabla \cdot c)^2\}$$

$$(A.7)$$

and

$$h = [C_p T]$$

# Cauchy-Riemann Equations

## Cartesian coordinates:

$$\partial \Phi/\partial x = \partial \psi/\partial y \tag{A.8}$$

$$\partial \Phi/\partial y = -\partial \psi/\partial x \tag{A.9}$$

$\Phi = $ velocity potential: $\psi = $ stream function

## Polar cylindrical coordinates:

$$\partial \Phi/\partial r = (1/r)(\partial \psi/\partial \theta) \tag{A.10}$$

$$(1/r)\partial \Phi/\partial \theta = -(\partial \psi/\partial r) \tag{A.11}$$

# Euler Turbine Equations

## Cartesian coordinates:

$$x\text{-component: } \rho[V_x(\partial V_x/\partial x) + V_y(\partial V_x/\partial y)] = -(\partial p/\partial x) + \rho g_x \qquad (A.12)$$

$$y\text{-component: } \rho[V_x(\partial V_y/\partial x) + V_y(\partial V_y/\partial y)] = -(\partial p/\partial y) + \rho g_y \qquad (A.13)$$

## Polar cylindrical coordinates:

$$r\text{-component: } \rho[V_r(\partial V_r/\partial r) + (V_\theta/r)(\partial V_r/\partial \theta) - (V_\theta^2/r)] = -(\partial p/\partial r) + \rho g_r \qquad (A.14)$$

$$\theta\text{-component: } \rho[V_r(\partial V_\theta/\partial r) + (V_\theta/r)(\partial V_\theta/\partial \theta) + (V_\theta V_r)/r] = -(1/r)(\partial p/\partial \theta) + \rho g_\theta \qquad (A.15)$$

# SPECIFIC GRAVITY AND VISCOSITY OF WATER AT ATMOSPHERIC PRESSURE

| Temperature °C | Specific Gravity | Absolute Viscosity $N\text{-}s/m^2 \times 10^3$ | Kinematic Viscosity $m^2/s \times 10^6$ |
|---|---|---|---|
| 0 | 0.9999 | 1.787 | 1.787 |
| 2 | 1.0000 | 1.671 | 1.671 |
| 4 | 1.0000 | 1.567 | 1.567 |
| 6 | 1.0000 | 1.472 | 1.472 |
| 8 | 0.9999 | 1.386 | 1.386 |
| 10 | 0.9997 | 1.307 | 1.307 |
| 12 | 0.9995 | 1.235 | 1.236 |
| 14 | 0.9998 | 1.169 | 1.170 |
| 16 | 0.9990 | 1.109 | 1.110 |
| 18 | 0.9986 | 1.053 | 1.054 |
| 20 | 0.9982 | 1.002 | 1.004 |
| 22 | 0.9978 | 0.9548 | 0.9569 |
| 24 | 0.9973 | 0.9111 | 0.9135 |
| 26 | 0.9968 | 0.8705 | 0.8732 |
| 28 | 0.9963 | 0.8327 | 0.8358 |
| 30 | 0.9957 | 0.7975 | 0.8009 |
| 32 | 0.9951 | 0.7647 | 0.7685 |
| 34 | 0.9944 | 0.7340 | 0.7381 |
| 36 | 0.9937 | 0.7052 | 0.7097 |
| 38 | 0.9930 | 0.6783 | 0.6831 |
| 40 | 0.9923 | 0.6529 | 0.6580 |
| 42 | 0.9915 | 0.6291 | 0.6345 |
| 44 | 0.9907 | 0.6067 | 0.6124 |

| Temperature °C | Specific Gravity | Absolute Viscosity N-s/m$^2$ × 10$^3$ | Kinematic Viscosity m$^2$/s × 10$^6$ |
|---|---|---|---|
| 46 | 0.9898 | 0.5856 | 0.5916 |
| 48 | 0.9890 | 0.5656 | 0.5719 |
| 50 | 0.9881 | 0.5468 | 0.5534 |
| 52 | 0.9871 | 0.5390 | 0.5359 |
| 54 | 0.9862 | 0.5121 | 0.5193 |
| 56 | 0.9852 | 0.4961 | 0.5036 |
| 58 | 0.9842 | 0.4809 | 0.4886 |
| 60 | 0.9832 | 0.4665 | 0.4745 |
| 62 | 0.9822 | 0.4528 | 0.4610 |
| 64 | 0.9811 | 0.4398 | 0.4483 |
| 66 | 0.9800 | 0.4273 | 0.4360 |
| 68 | 0.9789 | 0.4155 | 0.4245 |
| 70 | 0.9778 | 0.4042 | 0.4134 |
| 72 | 0.9766 | 0.3934 | 0.4028 |
| 74 | 0.9755 | 0.3831 | 0.3927 |
| 76 | 0.9743 | 0.3732 | 0.3830 |
| 78 | 0.9731 | 0.3638 | 0.3738 |
| 80 | 0.9718 | 0.3537 | 0.3640 |
| 82 | 0.9706 | 0.3460 | 0.3565 |
| 84 | 0.9693 | 0.3377 | 0.3484 |
| 86 | 0.9680 | 0.3297 | 0.3406 |
| 88 | 0.9667 | 0.3221 | 0.3332 |
| 90 | 0.9653 | 0.3147 | 0.3260 |
| 92 | 0.9640 | 0.3076 | 0.3191 |
| 94 | 0.9626 | 0.3008 | 0.3125 |
| 96 | 0.9612 | 0.2942 | 0.3061 |
| 98 | 0.9584 | 0.2879 | 0.3000 |
| 100 | 0.9584 | 0.2818 | 0.2940 |

# VAPOR PRESSURE CHART FOR VARIOUS LIQUIDS

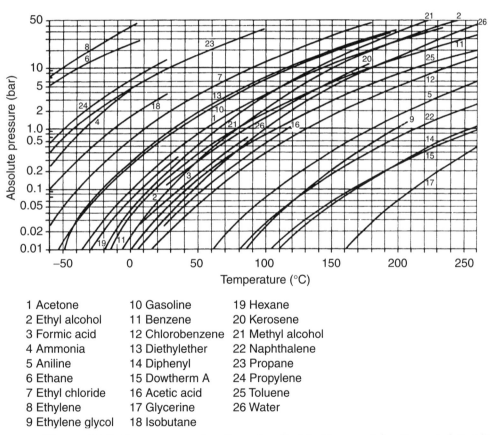

**Figure A3.1** Vapor pressure chart for various liquids. (Courtesy Sulzer Pumps Ltd, Zurich)

1 Acetone
2 Ethyl alcohol
3 Formic acid
4 Ammonia
5 Aniline
6 Ethane
7 Ethyl chloride
8 Ethylene
9 Ethylene glycol
10 Gasoline
11 Benzene
12 Chlorobenzene
13 Diethylether
14 Diphenyl
15 Dowtherm A
16 Acetic acid
17 Glycerine
18 Isobutane
19 Hexane
20 Kerosene
21 Methyl alcohol
22 Naphthalene
23 Propane
24 Propylene
25 Toluene
26 Water

# DENSITIES OF VARIOUS LIQUIDS

| Name | Temp. °C | Sp. Gr. | Name | Temp. °C | Sp. Gr. |
|---|---|---|---|---|---|
| Gasoline | | | Seawater | 15 | 1.02–1.03 |
| aviation | 15 | 0.72 | Mineral lubricating oil | 20 | 0.88–0.96 |
| normal | 15 | 0.72–0.74 | Naphthalene | 19 | 0.76 |
| Diesel fuel | 15 | 0.82–0.84 | Paraffin oil | 20 | 0.90–1.02 |
| Gear oil | 15 | 0.92 | Crude oil | | |
| Fuel oil | | | Arabian | 20 | 0.85 |
| light | 15 | 0.86–0.91 | Iranian | 20 | 0.835 |
| medium | 15 | 0.92–0.99 | Kuwaiti | 20 | 0.87 |
| bunker C | 15 | 0.95–1.0 | Trinidad | 20 | 0.885 |
| Hydraulic oil | 20 | 0.875 | Venezuelan | 20 | 0.935 |
| Sugar solution | | | Silicone oil | 20 | 0.94 |
| 10% | 20 | 1.04 | Bituminous coal tar oil | 20 | 0.9–1.1 |
| 20% | 20 | 1.08 | Vegetable oils | 15 | 0.090–0.97 |
| 40% | 20 | 1.18 | Machine oil | | |
| 60% | 20 | 1.28 | light | 15 | 0.88–0.90 |
| Kerosene | 15 | 0.78–0.82 | medium | 15 | 0.91–0.935 |

# MATHEMATICAL AND PHYSICAL CONSTANTS

**Mathematical constants:**

$$e = 2.71828\ldots$$
$$\pi = 3.14159\ldots$$

**Universal gas constant:**

$$R = 8.31451 \text{ kJ/(kg-mol)(K)}$$
$$= 8.31451 \text{ m}^3\text{-Pa/(kg-mol)(K)}$$
$$= 1545.36 \text{ ft-lb}_f/(\text{s}^2)(\text{lb-mol})(\text{R})$$
$$= 4.968 \times 10^4 \text{ lb}_m\text{-ft}^2/(\text{s}^2)(\text{lb-mol})(\text{R})$$

**Acceleration due to gravity:**

$$g = 9.80665 \text{ m/s}^2$$
$$= 32.174 \text{ ft/s}^2$$

# CONVERSION FACTORS

Area: $1 \text{ ft}^2 = 9.2903 \times 10^{-2} \text{ m}^2$

Density: $1 \text{ lb}_m/\text{ft}^3 = 16.0186 \text{ kg/m}^3$

Energy: $1 \text{ ft} - \text{lb}_f = 1.3558 \text{ J}$

$1 \text{ BTU} = 1.0551 \times 10^3 \text{ J}$

Force: $1 \text{ lb}_f = 4.4482 \text{ N}$

Length: $1 \text{ ft} = 0.30480 \text{ m} = 30.48 \text{ cm}$

Mass: $1 \text{ lb}_m = 4.536 \times 10^{-1} \text{ kg} = 453.6 \text{ g}$

Mass flow rate: $1 \text{ lb}_m/\text{h} = 1.2600 \times 10^{-4} \text{ kg/s}$

**Power:**

$1 \text{ ft-lb}_f/\text{s} = 1.3558 \text{ W}$

$1 \text{ bhp} = 745.7 \text{ W}$

$1 \text{ BTU/min} = 1.7584 \times 10^{-1} \text{ W}$

$1 \text{ cal/s} = 4.1840 \text{ W}$

**Pressure:**

$1 \text{ atm} = 1.0133 \text{ Pa} = 101.33 \text{ kPa}$

$1 \text{ psi} = 6.8948 \times 10^3 \text{ Pa}$

$1 \text{ bar} = 1 \times 10^5 \text{ Pa}$

$$1 \text{ mm Hg} = 133.32 \text{ Pa}$$

$$1 \text{ ft water} = 2988.9 \text{ Pa} = 2.9889 \text{ kPa}$$

$$1 \text{ m water} = 9806.38 \text{ Pa} = 9.80638 \text{ kPa}$$

## Temperature:

$$°F = (1.8) \times °C + 32$$

$$K = °C + 273.15$$

## Velocity:

$$1 \text{ ft/s} = 0.3048 \text{ m/s}$$

$$1 \text{ mph} = 0.44704 \text{ m/s}$$

$$1 \text{ kph} = 2.7777 \times 10^{-1} \text{ m/s}$$

## Viscosity (absolute):

$$1 \text{ cp} = 10^{-3} \text{ Pa-s}$$

$$1 \text{ lb}_f\text{-s/ft}^2 = 4.787 \times 10^{-1} \text{ Pa-s}$$

## Viscosity (kinematic):

$$1 \text{ ft}^2/s = 9.2903 \times 10^{-2} \text{ m}^2/s$$

$$1 \text{ cs} = 1 \times 10^{-6} \text{ m}^2/s$$

## Volume:

$$1 \text{ ft}^3 = 2.8317 \times 10^{-2} \text{ m}^3$$

$$1 \text{ liter} = 1 \times 10^{-3} \text{ m}^3$$

$$1 \text{ Imperial (English) gallon} = 4.5460 \times 10^{-3} \text{ m}^3$$

$$1 \text{ U.S. gallon} = 3.7853 \times 10^{-3} \text{ m}^3$$

**Volumetric flow rate:**

$$1 \text{ ft}^3/\text{min} = 4.7195 \times 10^{-4} \text{ m}^3/\text{s}$$

$$1 \text{ Imperial gallon}/\text{min} = 7.5766 \times 10^{-5} \text{ m}^3/\text{s}$$

$$1 \text{ U.S. gallon}/\text{min} = 6.3089 \times 10^{-5} \text{ m}^3/\text{s}$$

# BEAM FORMULAS AND FIGURES

[Refer to accompanying Figures A7.1 to A7.6]

| Beam Type | Reaction, R | Bending Moment | Deflection, y |
|---|---|---|---|
| A7.1 | $R_1 = Fa/L$:<br>$R_2 = Fb/L$ | $M_C = Fab/L$ | $y_{max} = (Fab)(a+2b)[3a(a+2b)]^{0.5} \div 27EIL$ |
| A7.2 | $R_1 = R_2 = F/2$ | $M_C = FL/8$ | $y_C = (5/384)(FL^3/EI)$ |
| A7.3 | $R_1 = R_2 = F/2$ | $M_B = M_C = Fa/2$ | $y_A = Fa^2(3L - 4a)/12EI$<br>$y_B = Fa(L - 2a)^2/16EI$ |
| A7.4 | $R_1 = R_2 = F/2$ | $M_A = M_C = Fb/2$ | $y_C = (Fb/12EI)(0.75\ L^2 - b^2)$ |
| A7.5 | $R_1 = R_2 = F/2$ | $M_C = FL/4$ | $y_C = (1/48)(FL^3/EI)$ |
| A7.6 | $R_1 = R_2 = F/2$ | $M_C = (F/8)(2b+1)$ | $y_C = [(5 - 24b^2 + 16b^4)$<br>$(FL^3)] \div [384(1 - 2b)EI]$ |

E = modulus of elasticity
I = moment of inertia

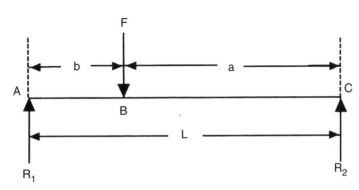

**Figure A7.1** Simply supported beam, single load at distance b from one end.

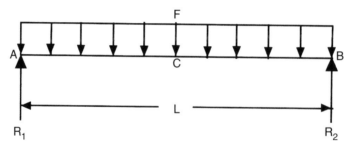

**Figure A7.2** Simply supported beam, uniform load.

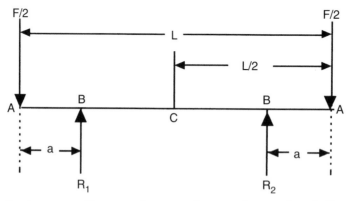

**Figure A7.3** Simply supported beam, each support distance a from each end of beam. Equal loads at each end of the beam.

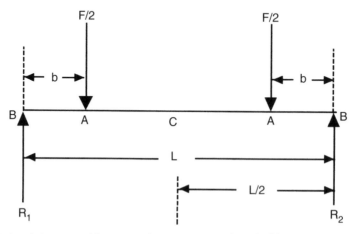

**Figure A7.4** Simply supported beam, each support at each end of beam. Equal loads distance b from the ends of the beam.

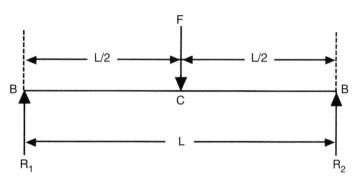

**Figure A7.5** Simply supported beam, each support at each end of beam, single load at the middle of the beam.

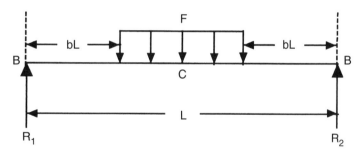

**Figure A7.6** Simply supported beam, each support at each end of beam, uniform load distributed in the middle of the beam, distance bL from each end.

# CHARTS FOR FLOWS
# THROUGH FITTINGS

The following charts give the additional flow resistances resulting from the presence of various fittings, bends, valves, expansions, and contractions in a pipe system.

In Figure A8-1, the head loss in a bend is given by:

$$H = K_b \left( \frac{1}{2g} \rho U^2 \right) \tag{A8.1}$$

Figure A8-2 relates the Thoma cavitation number, $\sigma$, to the geometry of the bend.

Figure A8-3 shows the total head loss in terms of the geometry of miter bends and the equivalent length of pipe measured in pipe diameters.

Figure A8-4 shows the resistance of enlargements and contractions in terms of head loss coefficient and geometry. The relationships are shown in the inset diagrams.

Figure A8-5 shows the head loss coefficients for a series of pipe entrances and exits.

Figure A8-6 shows loss coefficients for fittings.

Figure A8-7 shows flow coefficients, K as a function of Reynolds number, Re for orifices.

Figure A8-8 shows the dimensional parameters in the inset diagram for losses in diffusers. AR refers to the area ratio of the exit pipe to the inlet pipe, that is, $A_2/A_1$. $C_p$ on the curves is the pressure recovery coefficient, that is,

$$C_p = (\text{static pressure recovery})/ \left( \frac{1}{2} \rho U_1^2 \right) \tag{A8.2}$$

**324**

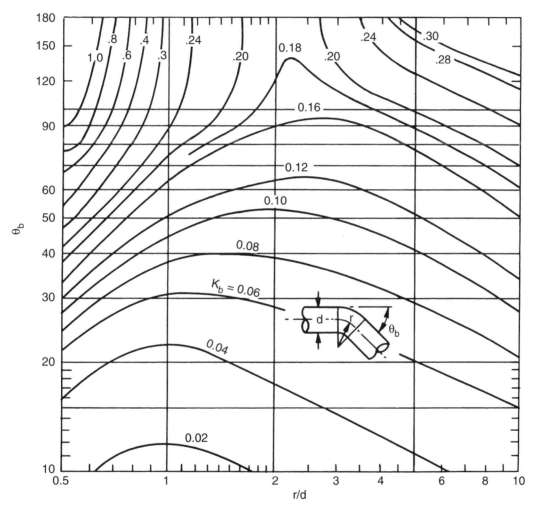

**Figure A8.1** Bend performance chart. (Courtesy BH Group Ltd.)

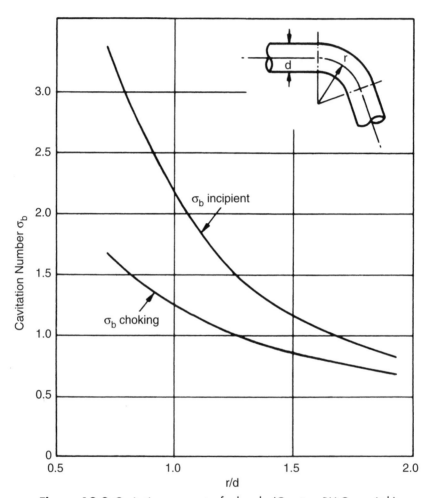

**Figure A8.2** Cavitation parameter for bends. (Courtesy BH Group Ltd.)

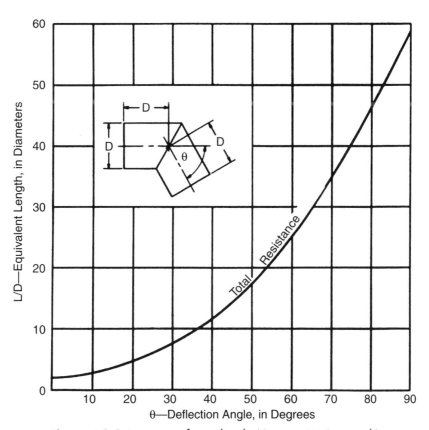

**Figure A8.3** Resistance of miter bends. (Courtesy BH Group Ltd.)

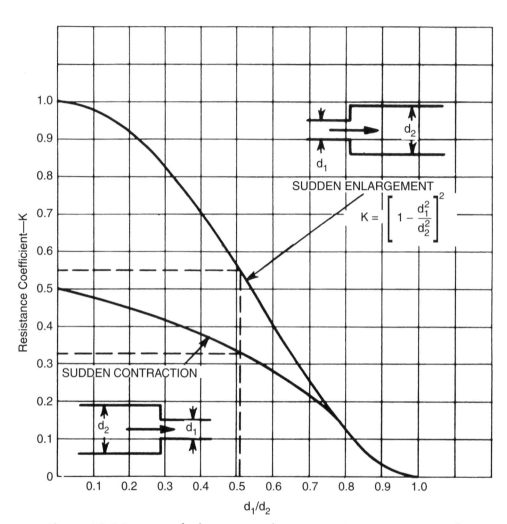

**Figure A8.4** Resistance of enlargements and contractions. (Courtesy BH Group Ltd.)

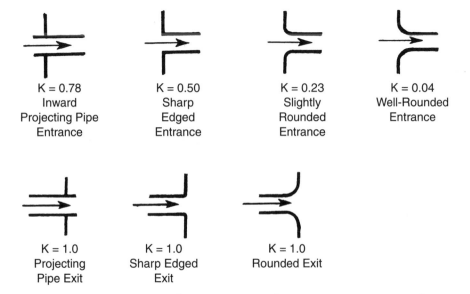

**Figure A8.5** Resistance of pipe entrances and exits. (Courtesy BH Group Ltd.)

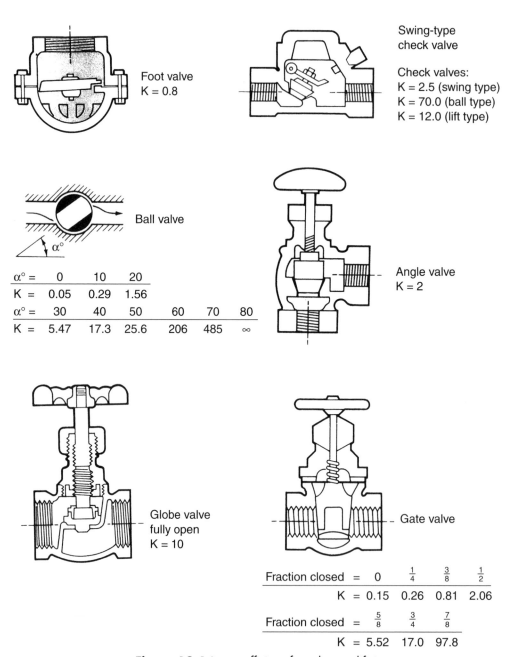

**Figure A8.6** Loss coefficients for valves and fittings.

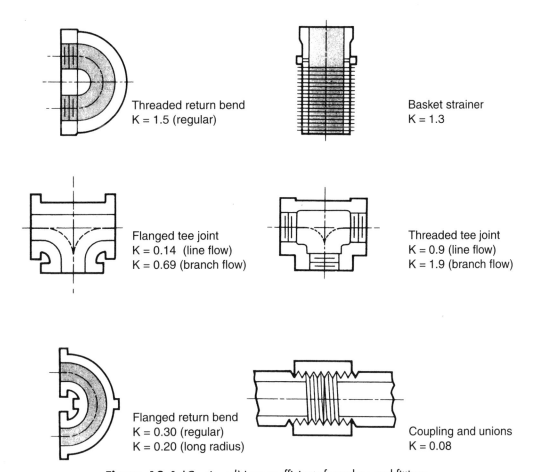

Threaded return bend
K = 1.5 (regular)

Basket strainer
K = 1.3

Flanged tee joint
K = 0.14 (line flow)
K = 0.69 (branch flow)

Threaded tee joint
K = 0.9 (line flow)
K = 1.9 (branch flow)

Flanged return bend
K = 0.30 (regular)
K = 0.20 (long radius)

Coupling and unions
K = 0.08

**Figure A8.6** (*Continued*) Loss coefficients for valves and fittings.

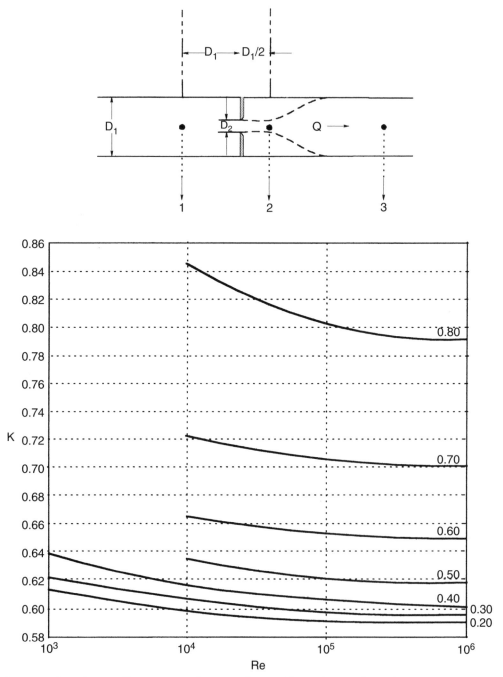

**Figure A8.7** Flow coefficient, K as a function of Reynolds number, Re for orifices. Parameter: β = diameter ratio = $D_0/D_1$. (Courtesy ASME—American Society of Mechanical Engineers)

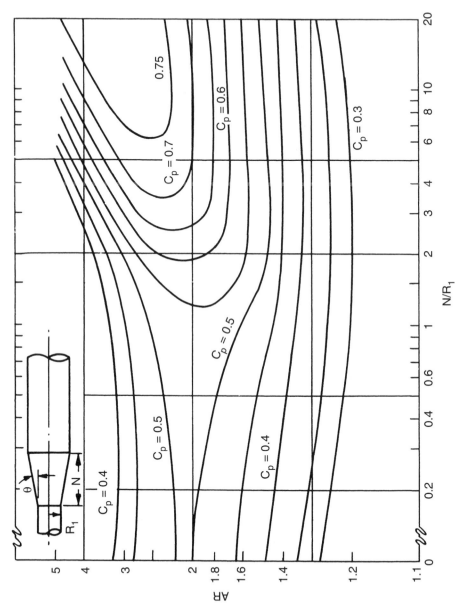

**Figure A8.8** Diffuser performance chart. (Courtesy BH Group Ltd.)

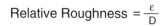

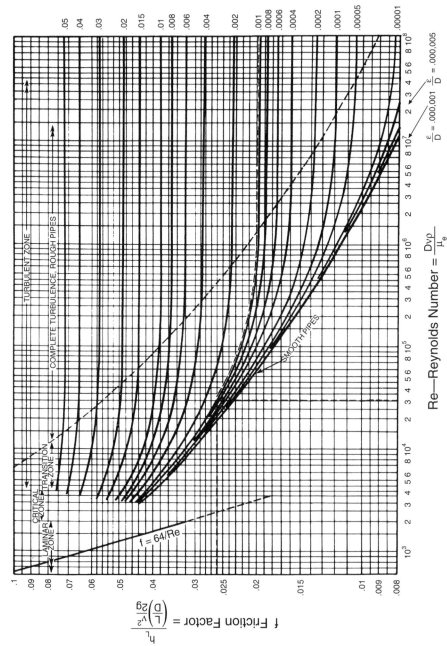

**Figure A9.1** Friction factors for fully developed pipe flow. (Taken from Moody, L.E., "Friction factors for pipe flow." *Trans. ASME* (1944), pp. 671–684. Courtesy ASME)

# VALUES OF PIPE ROUGHNESS, $\varepsilon$ FOR VARIOUS MATERIALS

| Material | $\varepsilon$ (cm) |
|---|---|
| Riveted steel | 0.09–0.9 |
| Concrete | 0.03–0.3 |
| Wood stave | 0.018–0.09 |
| Cast iron | 0.025 |
| Galvanized metal | 0.015 |
| Asphalted cast iron | 0.012 |
| Commercial steel or wrought iron | 0.0046 |
| Drawn tubing | 0.00015 |

# CHARACTERISTIC VALUES OF WATER IN THE SATURATION STATES

**Figure A11.1** Characteristic values of water in the saturation state.
$\theta$ = Temperature °C: p = pressure: $\rho$ = density;
$c_p$ = specific heat at constant pressure: $\eta$ = dynamic viscosity;
$\nu$ = kinematic viscosity.

| $\theta$° C | p bar | $\rho$ kg/m³ | $c_p$ kJ/kg · K | $\eta$ $10^{-6}$ kg/m · s | $\nu$ $10^{-6}$ m²/s |
|---|---|---|---|---|---|
| 0.01 | 0.006112 | 999.8 | 4.217 | 1750 | 1.75 |
| 10 | 0.012271 | 999.7 | 4.193 | 1300 | 1.30 |
| 20 | 0.023368 | 998.3 | 4.182 | 1000 | 1.00 |
| 30 | 0.042417 | 995.7 | 4.179 | 797 | 0.800 |
| 40 | 0.073749 | 992.3 | 4.179 | 651 | 0.656 |
| 50 | 0.12334 | 988.0 | 4.181 | 544 | 0.551 |
| 60 | 0.19919 | 983.2 | 4.185 | 463 | 0.471 |
| 70 | 0.31161 | 977.7 | 4.190 | 400 | 0.409 |
| 80 | 0.47359 | 971.6 | 4.197 | 351 | 0.361 |
| 90 | 0.70108 | 965.2 | 4.205 | 311 | 0.322 |
| 100 | 1.0132 | 958.1 | 4.216 | 279 | 0.291 |
| 110 | 1.4326 | 950.7 | 4.229 | 252 | 0.265 |
| 120 | 1.9854 | 942.9 | 4.245 | 230 | 0.244 |
| 130 | 2.7012 | 934.6 | 4.263 | 211 | 0.226 |
| 140 | 3.6136 | 925.8 | 4.258 | 195 | 0.211 |
| 150 | 4.7597 | 916.8 | 4.310 | 181 | 0.197 |
| 160 | 6.1804 | 907.3 | 4.339 | 169 | 0.186 |
| 170 | 7.9202 | 897.3 | 4.371 | 159 | 0.177 |
| 180 | 10.003 | 886.9 | 4.408 | 149 | 0.168 |

| $\theta°$ C | p bar | $\rho$ kg/m$^3$ | $c_p$ kJ/kg $\cdot$ K | $\eta$ $10^{-6}$ kg/m $\cdot$ s | $v$ $10^{-6}$ m$^2$/s |
|---|---|---|---|---|---|
| 190 | 12.552 | 876.0 | 4.449 | 141 | 0.161 |
| 200 | 15.551 | 864.7 | 4.497 | 134 | 0.155 |
| 210 | 19.080 | 852.8 | 4.551 | 127 | 0.149 |
| 220 | 23.201 | 840.3 | 4.614 | 122 | 0.145 |
| 230 | 27.979 | 827.3 | 4.686 | 116 | 0.140 |
| 240 | 33.480 | 813.6 | 4.770 | 111 | 0.136 |
| 250 | 39.776 | 799.2 | 4.869 | 107 | 0.134 |
| 260 | 46.940 | 783.9 | 4.986 | 103 | 0.131 |
| 270 | 55.051 | 767.8 | 5.126 | 99.4 | 0.129 |
| 280 | 64.191 | 750.5 | 5.296 | 96.1 | 0.128 |
| 290 | 74.448 | 732.1 | 5.507 | 93.0 | 0.127 |
| 300 | 85.917 | 712.2 | 5.773 | 90.1 | 0.127 |
| 310 | 98.697 | 690.6 | 6.120 | 86.5 | 0.125 |
| 320 | 112.90 | 666.9 | 6.586 | 83.0 | 0.124 |
| 330 | 128.65 | 640.5 | 7.248 | 79.4 | 0.124 |
| 340 | 146.08 | 610.3 | 8.270 | 75.4 | 0.124 |
| 350 | 165.37 | 574.5 | 10.08 | 70.9 | 0.123 |
| 360 | 186.74 | 528.3 | 14.99 | 65.3 | 0.124 |
| 370 | 210.53 | 448.3 | 53.92 | 56.0 | 0.125 |
| 374.15 | 221.20 | 315.5 | $\infty$ | 45.0 | 0.143 |

# INDEX